Nelson Yuan-sheng Kiang

Cortical Memory Functions

C. M. Fair

Cortical Memory Functions

1992
Birkhäuser
Boston · Basel · Berlin

C. M. Fair
Wakefield, RI 02879

Library of Congress Cataloging-in-Publication Data

Fair, Charles M.
Cortical memory functions / by C.M. Fair
p. cm.
Includes bibliographical references and indexes.
ISBN 0-8176-3615-3 (H : alk. paper). -- ISBN 3-7643-3615-3 (H : alk. paper)
1. Memory. 2. Neocortex. I. Title.
[DNLM: 1. Cerebral Cortex--physiology. 2. Memory--physiology. 3. Neurophysiology. WL 307 F 163c]
QP406.F33 1992
612.8'2--dc20
DNLM/DLC 92-2553
for Library of Congress CIP

Printed on acid-free paper.

ISBN 0-8176-3615-3
ISBN 3-7643-3615-3

Typset by ARK Publications, Inc., Newton Centre, MA
Printed and bound by Quinn-Woodbine, Woodbine, NJ
Printed in the U.S.A.

9 8 7 6 5 4 3 2 1

When memory and anticipation are completely absent, there is complete conformity to the average influence of the immediate past.... Such a situation produces the activity of mere matter. When there is memory, however feeble and short-lived, the average influence of the immediate past... ceases to dominate completely.... Thus the universe is material in proportion to the restriction of memory and anticipation.

According to this account... there is no need to postulate two essentially different types of active entities, namely the purely material... and the entities alive with various modes of experiencing.... This conclusion has the advantage of indicating the possibility of the emergence of life from the lifeless material of this planet—namely, by the gradual emergence of memory and anticipation.

A. N. Whitehead (1949)

Contents

Foreword by the Author

A potential reader, seeing a book with a title such as this one, will have two principal questions about it:

(1) Does the announced subject warrant a book-length treatment at this time?

(2) Is the subject of the book close enough to the reader's or journal's special interests to warrant a serious reading or a review?

This book is about long-term memory—more precisely, about the way in which short- and long-term memory functions are interwoven in neocortex. It is about the structure, including fine-structure, of the memories underlying our generic or specific recognition of everyday data of sense.

To answer the first question above: The subject does warrant full-length treatment at this time because (as I will try to show) we have accumulated sufficient neurophysiological, biochemical, neuroanatomic, psychophysical, and psychological evidence to make a comprehensive model of cortical memory-functions possible.

To answer the second question: If the interests of a reader are stringently confined to some aspect of central nervous function, he may not consider this work worth the trouble. And those who do will need either to consult the sources I have cited from fields other than their own, or to take it on faith that I have correctly interpreted the data cited.

In the present climate of opinion, I am aware that this may be asking a bit much. For example, I have the (perhaps incorrect) impression that neuroanatomy today has become a specialty unto itself. Its practitioners draw conclusions as to the functional significance of their findings very sparingly, while neuroscientists who are primarily concerned with some aspect of function tend to have a knowledge of neuroanatomy that is limited by their special interest and, even then, is often somewhat schematic.

This means that many neuroanatomists may balk at the functional interpretation I have given their sort of data, whereas many neuroscientists in other fields may balk at references to neuroanatomy they cannot confirm from memory and begrudge the time to look up. Members of both groups may consequently decline to read the book on the ground that it is too risky or too difficult to follow, or both.

Let me say, in my own defence, that although the argument presented here draws upon data from a number of different fields, they fit together in ways that are often quite striking; that the problem of memory is in itself an important one; and that theory in neuroscience, if it is to fulfill its function of unifying our understanding by way of testable propositions, has got, at some point, to become interdisciplinary.

I submit that the present theory meets those requirements and may therefore justify the effort it asks of the reader. To the extent that it passes the critical test of experiment, it will also give research in the fields related to memory a clearer rationale, thereby, perhaps, saving time.

Introduction

What follows is an outline of the theory of memory functions developed in this monograph. It will not include the supporting data and references used throughout in the text, but will simply present the essentials of the argument.

There are at least three levels to the problem of memory:

Level 1 concerns the changes by which single units come to "remember" (respond preferentially to) certain of their inputs.

Level 2 concerns the way in which these unitary changes, occurring selectively in certain assemblies in six-layered cortex, give rise to the several known forms of memory.

Level 3 concerns the problem of translation. Any theory of memory must be able to show that these physiological and neuroanatomic substrates have clear psychological correlates.

Level 2: The Areal Distribution of Memory Functions

Let me begin with a Level 2 problem—that of the distribution of long-term memory functions in posterior neocortex. The type of long-term memory referred to is (1) consciously available in humans, and (2) "declarative" rather than procedural (having to do with recalled facts rather than learned skills).

In man, and apparently in some other species, the formation of indefinitely long-lasting memories appears to depend upon the hippocampus. Once established, such memories survive bilateral removal of the hippocampus. Following bilateral hippocampal ablation, however, no further additions to long-term memory (LTM) can be made. Thus, the hippocampus figures in LTM formation, but the resulting memories are laid down elsewhere.

Other data suggest that they are laid down in neocortex. For example, sufficient bilateral damage to cortex of the fusiform and lingual gyri or their white matter connections can result in a generalized form of prosopagnosia—a failure of specific visual recognition.

Such patients can still generically identify a visual input; they have simply become unable to recognize a particular face or object or animal as one previously known to them. The implication is that they have lost, or lost access to, the long-term memories making such specific visual identification possible.

Cortex of the lingual and fusiform gyri (Figure 3-1 in the text) is mesial. It adjoins the parahippocampal gyrus; and (if one estimates the distance in synapses in the transcortical relay) it lies at least one synapse further away from the primary visual receiving area than does cortex responsible for generic recognition. This is shown by the fact that if the lesions extend bilaterally into the latter area (i.e., towards primary visual cortex) generic recognition is also impaired, though the patient is not blind.

In other words, generic (visual) memories are laid down first, and specific memories second, or at a later stage in the cortical relay.

There are two other points to note. If the cortical memory systems are sequentially organized in this way, our perceptual processes must follow the same sequence. It has been reported that they do—that, on a scale of milliseconds, our generic recognition of a given input precedes specific. (The lag time, though short, is sometimes subjectively perceptible.)

The second point is that our generic awareness of our surroundings appears to be acquired without much conscious effort, developing simply as a by-product of our day-to-day exposure to the world. In vision alone, the number of inputs we become able to identify in this way—by type—is quite large. Furthermore, this automatic classification system is quite a complex one, consisting of overlapping hierarchies such that, for instance, dogs and cars form a class ("things that can move") but also belong to other distinct classes (e.g., living vs. nonliving things).

The generic memory system must of necessity be quite large—must take up a good deal of cortical space—to accommodate this degree of elaboration. In fact, that is the case. For while this memory system cannot of itself support specific recognition, many of its components, as we shall see, form part of the pattern equivalent to the memory of a particular person, event, or object. That is, our specific long-term memories are constructed from generalized components, and owe their specificity to the patterns formed by these.

In contrast to the number of inputs we become able to identify generically, the number we recognize specifically is much smaller. And whereas the memories underlying generic recognition take shape automatically as a by-product of daily experience, those underlying specific recognition are formed selectively or (crudely put) require extra work. In the central nervous system, that additional work is supplied by processes equivalent to what we call motivation (which often or usually involves corresponding feeling-states or what we call emotion). In other words, what we select for long-term recall will generally have some associative connection with our own most basic concerns.

To the extent that we are motivated or have a "drive" to do or remember something, our metabolic rate tends to go up, transaction rates in the central nervous system tend to rise accordingly, our foci of attention tend to become sharper, and so on. It is during the rising and declining phases of such episodes of central arousal that the processes leading to long-term memory formation are set in motion. It is in this sense that the formation of memories representing specific persons, events, objects, etc., is more episodic and also more energy intensive than is the formation of memories mediating generic recognition of the same sorts of data.

Given that these may be the conditions leading to long-term memory formation, how are they set up in the cortex? The answer is: by way of input from the hippocampus, supported by some concurrent rise in nonspecific (reticular) arousal.

(As will be seen in the text, this nonspecific input is of some importance, since it affects both the hippocampus and the cortex, and leads in the latter to "coactivation,"—a convergence effect now considered in some way necessary to memory formation but not as yet clearly understood.)

It is known that the hippocampal system receives projections funneled into it from many cortical association areas. It also sends return projections to most or all of these same areas. The conclusion suggested by the two findings with which we began is that the selective process underlying long-term memory formation must depend upon these circular connections, for when they are cut by bilateral destruction of the hippocampus, humans (and evidently some animals) become unable to add further to LTM.

Let me briefly explain why the hippocampus may play this role in cortical memory functions, and suggest in more detail what that role may be.

The hippocampus receives input from the ventral tegmental area, a part of the brainstem known to be of major importance in "motivation" or behavior arising out of basal adaptive drive-states. And more than other forebrain structures, the hippocampus is responsive to circulatory levels of cortisol or ACTH. These inputs (to name only a few) make it a sensitive monitor of the physiological status quo, and regulate its output to the cortex accordingly.

By way of its two-way cortical connections, the hippocampus is able to recycle immediate or precurrent sensory data over several loops, in this way producing what I have called "circular reinforcement." This amounts to a form of automatic internal rehearsal, similar in principle to the conscious mental rehearsal we sometimes use to be sure we remember a telephone number. In proportion to the degree that it is activated by the other (extracortical) inputs just mentioned, the hippocampus is able to extend the duration of short-term memory (STM).

This, indeed, is its primary function—to recycle, and thereby put on "hold" (in the cortex), sense-data that may prove to be adaptively important. Many of these data are, in fact, of no lasting adaptive significance and do not go on to become incorporated into LTM.

Long-term memories result from the same process of circular reinforcement only when our "motivational" state and its accompanying central arousal exceed some minimum of intensity and/or duration. (There is a tradeoff relation between the two.) When that minimum is exceeded, recirculation of data from the cortex by way of the hippocampus results in their "consolidation"—incorporation into long-term memory.

In other words, via its bidirectional connections with the neocortex, the hippocampus acts to select many items of sensory experience for *potential* inclusion in LTM. The half-life of the items in STM is thereby extended (as in the "primacy" effect in serial learning tests), but in most cases the process stops there. The items are not lastingly recalled.

(As evidence for this relationship, I should mention that after bilateral hippocampectomy there is no primacy effect, and the duration of STM becomes in general more rigidly limited, because the system normally responsible for extending it has been eliminated.)

In effect, then, LTM works by selections made from (hippocampally extended) STM, on a many-are-called-but-few-are-chosen principle. The key question is why the items chosen should differ so dramatically in duration from those that are not. What is it, during consolidation, that causes extended short-term memories to become indefinitely long-lasting? Since the hippocampus apparently figures in both types of memory formation, how do we account for the difference between them?

To explain it, I have proposed that the hippocampus acts in two stages on the cortical memory systems. In the first, the recycling of sense-data (cortex → hippocampus → cortex) results in their dynamic maintenance in the cortical registers. This is the "hold" function that acts, as mentioned, to lengthen STM. It depends upon continued activity over the circuits involved and is roughly analogous to RAM memory.

The second stage, or consolidation, involves the biochemical modification of a subset of the cortical units reached by hippocampal playback, and is time-dependent on a scale of minutes to hours. (Consolidation can be blocked by protein synthesis inhibitors. And if anesthetics are administered during consolidation, they have graded effects on retention, from maximal if given immediately post-training, to nil if given after a sufficient interval.)

I have supposed that during consolidation, return fibers from the hippocampal system may release trophic factors that result in long-lasting site-specific changes in the postsynaptic membrane of the cortical units to which they project. In the visual system, for example, these would be neurons of the fusiform or lingual gyri—the parts of cortex that in man appear to be necessary to specific visual recognition.

Level 1: The Unitary Mechanics Problem

This brings us to the first problem mentioned above—that of the way memory is represented at the level of single neurons. It also brings us to the most speculative feature of the proposed hypothesis, which concerns the unitary basis of long-term memory, and the features of hippocampal-cortical input that may make that form of memory unique.

The hippocampus is known to respond, during seizures, by producing the mRNAs of four immediate effect genes (IEGs), one of which is also known to figure in long-term potentiation—a phenomenon long thought to be in some way critical to memory formation.

The hippocampus is also a source of trophic factors (synapsins; BDNF), including one (NT-3) reported to have limited distribution elsewhere in the brain. It is still not known whether projections of the hippocampus or the subicular complex release such factors in the cortex, still less whether they activate immediate effect genes in the recipient neurons, or otherwise act to produce the site-specific changes in postsynaptic membrane proposed here.

In postulating such a mechanism, I am working from logical intuition. Because long-term "declarative" memory can be of indefinite duration, it seemed reasonable to suppose it might involve trophic factors and/or activation of IEG mRNAs, either or both of which might be a means to inducing differential changes in receptor membrane of the memory-forming neurons downstream. (Note, however, that these are heuristic propositions; they can be proved true or false by experimental methods now available.)

My reason for supposing that LTM depends upon changes at certain input sites in postsynaptic membrane is that it would provide a unitary basis for preferential responses to certain inputs. If such changes were to be made permanent by way of biochemically induced alterations in receptor membrane, the result in single units or assemblies would be to produce lasting preferential responses to a subset of their possible inputs. And this, in turn, would provide the basis for specific long-term recall, e.g., of visual inputs in the lingual or fusiform gyri.

This result would *not* follow if, as in sensitization in invertebrates or in long-term potentiation (LTP) in vertebrates, the primary site of change were presynaptic. For here, the net effect is likely to be an increase in the probability of a unit's "firing" (as has been shown for LTP) or in the amounts of transmitter released at the synapse (as has been shown for sensitization), or both.

In the neocortex, where a single neuron may project to N others ($N \gg 1$), the specificity of the input(s) inducing these presynaptic changes in a given unit, even if conserved on the input side, would tend to be lost in the output. The neuron would simply show an across-the-board increase in its output to N others. (It has been reported that sensitization can, in fact, have behavioral effects suggesting this type of diffuse downstream activation.)

In simpler nervous systems such as that of *Aplysia*, or in converging, genetically constrained pathways such as those in the mammalian spinal cord, presynaptic learning may suffice, since the "choice" of final output paths is limited and fixed. Any one of N inputs to the system can, through sensitization or LTP of the units involved, come to drive it more efficiently or at higher probability than the remaining ($N - 1$) inputs.

The loss of specificity—that is, of the information represented by the configuration of inputs that initially caused the unit(s) to become sensitized—will not matter, since the specificity of the downstream response is hard-wired and therefore guaranteed. (LTP, essentially a form of sensitization, has been shown to occur in the mammalian cord.)

Presynaptic learning alone may not suffice in complex systems such as vertebrate association cortex, where the choice of possible pathways is comparatively great—where parts of the system, even at maturity, remain "open" or subject to further plastic change. In systems of this type, the information not supplied by hard-wiring downstream, will need to be conserved by mechanisms that, starting close to the cortical sensory input sources, conserve it across synapses.

For this reason, the primary mode of memory formation in neocortex may be postsynaptic, although it includes a facilitatory presynaptic component that in certain assemblies takes the form of long-term potentiation. I wish to stress these theoretical points, since if tenable, they will go far towards clarification of the problem of memory, including that of the role of LTP (currently a matter of some dispute).

Let me now try to complete the model of cortical memory functions that results from these ideas. We have seen that part of visual cortex supports generic visual recognition, while more distal parts of that cortex—parts that lie closer to the hippocampal system—support specific (visual) recognition.

On a variety of grounds, I have inferred this arrangement to mean that the visual cortex supporting generic recognition does so by way of a system of memories that are probabilistic. In this cortex data arriving from the periphery are put through a succession of cross-connected filters that sort them for invariances—most probable primary features, most probable combinations of such features, and so on, up to some unspecified level of generality.

This is the system that in vision acts automatically to sort our inputs according to their nearest fit with experientially established categories. I have supposed that at the unitary level, the primary mechanism is postsynaptic—use-induced establishment of preferred responses. (In the text, I review evidence suggesting that a similar kind of generic memory formation occurs in the auditory and somatosensory systems.)

The chief respect in which LTM differs from memories formed in this way is that these generic memories are not subject to biochemical "fixation" and thus remain subject to further change as a result of experience. There are other important differences.

The first is that generic memories are formed forward, by intracortical or cortico-thalamo-cortical relay, from prime receiving areas. Specific memories or LTM are formed recursively, by playback to some but not all of the areas involved in "forward" relay and generic recall.

The playback is in all probability not literally that—not reverse point-for-point transmission. It is mediated, rather, by a system of feedback loops that cut into cortical "feedforward" pathways at a series of strategic points, in this way reinforcing by redundance certain of the data currently being put through. Such loops are known to exist. I have inferred that their function is that of a selective positive feedback system.

Because of its global monitoring functions, its extensive bidirectional connections with the cortex, and its susceptibility to prolonged after-activity, the hippocampus is uniquely situated to "decide" which sensory inputs shall be selected for reinforcement in this way. And as already described, it has two modes of action; the short-term, which serves for dynamic maintenance of certain data in the sensory registers; and the long-term, which involves trophically mediated biochemical changes affecting certain neurons or assemblies in the cortical loop(s).

The question then is—which units in which assemblies? We have already concluded that the generic memory system is an extensive one, involving much of the association cortex surrounding the sensory and parasensory areas. We have also concluded that the data relayed through that system are generalized in successive stages, with some loss, at each stage, of the constituent details of the data represented.

Finally, from data on the human visual system, we know that memories mediating specific recognition lie synaptically further from primary visual cortex than do the visual memories mediating generic recognition. We also know that return fibers from the hippocampal system, including entorhinalis and the subicular complex, reach areas of association cortex that appear, in fact, to form part of the generic memory system. (In occipital cortex, this would be Brodmann area 19.)

Level 3: Psychological Features of Memory that Derive from its Physiological Organization

Taken together, what these data imply is that long-term memories are fixated only in part, and that the fixated components consist of the most generalized or schematic components of the memory as a whole. The referents of these memories—equivalent to their more detailed associative structure—are represented in the pathways leading *to* them, the paths followed during their consolidation.

The memories that lie in these pathways are of the unfixated or generic type. That is, they remain subject to probabilistic revision via inputs from the periphery. However, they will also show some reinforcement-through-use as a result of their inclusion in the recursive pathways that figure initially in their consolidation or subsequently in their retrieval.

Like the recursive processes that establish long-term memories having this composite structure, retrieval can also be recursive. Unlike recognition, which runs forward or from the outside in, retrieval can begin with activation of the more distal fixated components of a given memory and work by retracing some of the pathways originally followed during its consolidation.

Many of the referents of the memory retrieved in this way are *not* fixated. That is, they consist of memories of the generic system that, in structure or associative connections (which are part of their structure), remain subject to continuing probabilistic change. The tendency is consequently for the retrieval of older memories to become more uncertain as time passes, and for the memories themselves to become more schematic or stripped of their original detail.

In other words, the composite structure of cortical long-term memory proposed here accounts for what, in the psychological literature, is described as a common feature of long-term memory as we actually experience it. Other such features, including the (nonpathological) changes in memory occurring in old age, are similarly exemplified in the text.

The fact that long-term memory has this odd structure can best be understood in terms of its ancestry.

The specific thalamocortical projection system and immediately outlying cortex are phylogenetically recent, which means that this cortex has, as it were, been injected into a much older cortical system whose primary relations were, and still are, with the limbic system—principally the amygdala and hippocampus. It is this newer cortical system, including the sensory, parasensory and proximal association areas, that mediates generic recognition.

The advantage of this arrangement is that it permits the organism to construct, from highly resolved or fragmented sensory data (e.g., as represented in primary visual cortex), an inner representation of the external world that matches its day-to-day experience. And since experience itself is unpredictable, the memories in this cortical system remain correspondingly open to revision, thereby permitting the organism to alter its internal representation of the world as circumstances may require.

By definition, the memories in this system cannot support processes equivalent to specific recognition. In the visual mode, they can, for example, represent "dog" but not "that (particular) dog." A memory system mediating specific recognition (e.g., of a particular dog known to be particularly ferocious) has obvious adaptive value, and in its usual co-optative way, evolution has provided one.

The key structure in that system is the hippocampus, whose primordial function in amphibians may have been to establish associative connections between olfactory inputs and concurrent input-data in other modalities. In other words, it may have mediated the specific recall of odors in a form that included certain of their nonolfactory referents. (The phylogeny of the hippocampus and its relations to the amygdala are discussed in detail in the final sections of the monograph.)

Olfactory memories, as they involve recognition of molecular species, are necessarily quite specific. (An animal's survival may depend upon its olfactory resolution—its ability to distinguish an odor meaning danger from one meaning food.) The capacity of the hippocampal system to fixate adaptively significant inputs may have arisen out of that necessity—as an adaptation by which highly specific and long-lasting olfactory memories became associatively linked to more labile components reflecting the organism's day-to-day or moment-to-moment inputs in other modalities.

Except that the data it serves to fixate have largely or entirely ceased to be olfactory and now involve other modalities instead, the functional relation of the hippocampus to mammalian neocortex may be essentially the same as it was to the dorsal general pallium in amphibians. In both systems, the ancient and the modern, a part of the memories represented may be subject to fixation, or highly specific, and the remainder of their associations or referents, comparatively fluid.

It is for this reason that we find projections of the hippocampal system woven back into neocortex as far as visual area 19. For it is by these recursive pathways that the labile referents of long-term memories become linked to their more enduring "core" components. The latter lie in ventral and mesial cortex, in range perhaps of direct projections from the subicular complex, and therefore subject to biochemically mediated fixation.

The resulting long-term memory system is one that François Jacob (1977) would describe as "tinkered up"—a structure 25% steel and 75% putty—a compromise between forms of learning that might set too early and too hard, and forms that might be unreliable because they never set at all.

As explained in later chapters here, this account of cortical memory functions has some remarkable implications for cognitive psychology. The most striking of these concern, not the LTM system, important though it is, but the more labile generic memory system with which it is interwoven.

The basic argument may be summarized as follows. Six-layered cortex, though it shows considerable cytoarchitectural variation from area to area, is essentially somewhat stereotyped. The same applies to the intracortical relay system, which over longer trajectories chiefly involves the larger pyramidal cells of lower layer III. Because of the convergence of input upon them, these cells can be inferred to have generalizing or integrating functions. In addition, a variety of evidence suggests that, in contrast to the parvocellular assemblies presynaptic to them, the magnocellular arrays of layers lower III and V are characteristically memory-forming.

Putting these features together, we can conclude that the system is a probabilistic one; that at each stage in the cortical relay it tends to retain a generalized version of its most frequently encountered patterns of input; and that the process as a whole is one of successive invariance-extraction. This arrangement has the following consequences.

First, it puts together the pre-analyzed sensory data, provided by the specific thalamocortical projection system, into the elementary type-objects of perception. From the latter, it then extracts second-order invariances representing the categories into which these type-objects fall. That is, it generalizes across type-objects, in effect grouping them according to their more salient common features.

In a third stage—again, by repetition of essentially the same operations—it extracts from the data representing these categories the invariances they exhibit *across* categories. Memories of this type are more generalized in that they represent features held in common by a variety of inputs that may be otherwise unrelated.

Because such memories take shape automatically, they lead to an automatic awareness of analogical resemblances between things that are often, in fact, quite dissimilar. As a result of this third stage in generic memory formation, we do not have to be taught to use or understand figures of speech. They arise directly out of the way we perceive the world or, more precisely, out of the cumulative residues of perception in memory.

When the cross-categorical features that support our use of metaphor or metonymy are processed yet again, for invariances amongst themselves, the resulting memory data are generalized to the point of being what we call abstract. That is, they are so removed from the specific data of sense from which they derive that, like the "x" and "y" of algebra, they can stand for virtually anything. Memories of this type provide the basis for concepts such as "circularity" or for the ideal objects of geometry.

In short, the generic memory system, simply by the way it is organized —simply by the repetition of the same operations a sufficient number of times—presents us with many of the raw materials of thought ready-made. It is perhaps only in the *extent* of these operations or (equivalently) in the amount of cortex available for their performance, that we differ from other higher vertebrates.

The final section has to do with the way in which the long-term memory system, interwoven with this one, gives rise to the capacity for self-awareness—that is, to specific knowledge of the self as one phenomenon among others—and to the not always self-preoccupied *Doppelgänger* called consciousness.

Chapter 1
The Corticocortical Relay System

Abstract. It is proposed that, in posterior neocortex, the interareal projections arising in magnocellular pyramidal arrays of lower layer III, and projecting to similar arrays in lower III elsewhere, form the principal links in a several-stage cortical memory system. This conclusion is fundamental to much that follows.

I will begin with a report that is of critical importance to the argument developed in this monograph. It concerns the intracortical pathways by which incoming sense-data are relayed from prime receiving to association areas. As will be seen, the structural features of that relay system, taken together with a variety of other evidence presented later here, have clear implications as to the way information is processed in neocortex, which in turn has clear implications as to the form in which sensory data are retained in memory.

In the macaque, Pons et al. (1987) demonstrated the "functional dependency" of somatosensory area II (SII) on input from ipsilateral areas 3a, 3b, 1 and 2 in postcentral cortex. The key findings in this study were the following.

Input to SII from postcentral cortex was primarily or exclusively by transcortical relay although, as the authors point out, SII "has long been thought to receive a major projection from the ventroposterior nucleus (VP), the principal somatic relay nucleus of the thalamus." Ablation of 3a, 3b, 1 and 2 produced "virtually total" retrograde degeneration in the "somatotopically appropriate zone" of VP ipsilaterally. Lesions to SII produced "no discernible" changes in VP. (The authors cite other studies indicating that VP "sends at most a sparse input to SII.")

Recordings across the postcentral cortex after "the entire SII region had been removed" also showed no detectable change in the responses of postcentral units to tactile stimulation. The authors conclude from this result that the functional dependency of SII on 3a, 3b, 1 and 2 is not reciprocal.

Finally, they report that "recent anatomical evidence has revealed that neurons in each of the cytoarchitectonic fields of the postcentral strip, *particularly pyramidal cells in layer III*, project densely onto layer IV and the lower part of layer III in SII." (Italics added.)

They add that "this type of connection, from layer III of one cortical area to layers IV and lower III of another, has been shown in the visual system to be the specific route by which sensory information is transmitted in serial fashion from earlier cortical stations to later ones. . . . Since this same laminar pattern of connections has been found to link SII and the granular and dysgranular fields of the insula, it is likely that . . . SII in turn activates the insular cortex in a sequence analogous to the striate-prestriate-temporal sequence in vision," or the sequence in which SII itself is activated from postcentral cortex.

The authors conclude that this pattern of interareal relay may in fact reflect a "common cortical plan for the processing of sensory information in the various sensory modalities."

Jones (1981) mentions that in prime receiving areas, thalamic afferents, besides terminating in granular IV, "extend well upward among the large deeply situated pyramidal cells of layer III." He adds that "in areas outside the primary sensory areas (constituting some 80% of the primate cortex) they avoid layer IV and terminate mainly in the deep part of layer III." In other words, in association cortex, thalamocortical afferents access the same assemblies as do many of the longer interareal fiber systems. These neuroanatomic relations quite clearly exemplify the mixture of serial and parallel processing often said to be found in neocortex. In the somatosensory system described by Pons et al., the form of the relay appears to be serial only.

In the older literature, layer IV was described as very prominent in 17 and postcentral cortex, somewhat thinner in 41–42, but noticeably reduced in association cortex, in some areas showing an admixture of larger cells. In contrast, the arrays of larger pyramids in lower III and V were reported to become more prominent as one proceeded from koniose to parakoniose (sensory and parasensory) to generalized eulaminate (association) cortex. The detailed data compiled by Bailey and von Bonin (1951) suggest that, although these areal changes in laminar structure show considerable local variation, the trend of the change across the cortex is as just described.

Figure 1-1 summarizes some of the foregoing points. Note that interareal relay over longer fiber tracts (including the callosum) chiefly depends upon layer III, with some interareal fibers possibly originating in V; and that subcortical relay, e.g., to the structures shown in the upper panel, depends chiefly upon V. Thus V–VI, the layers making up what Lorente de Nó called the "internal lamina" (1951), are responsible for columnar final outflows, in the sense that many of these leave the cortex.

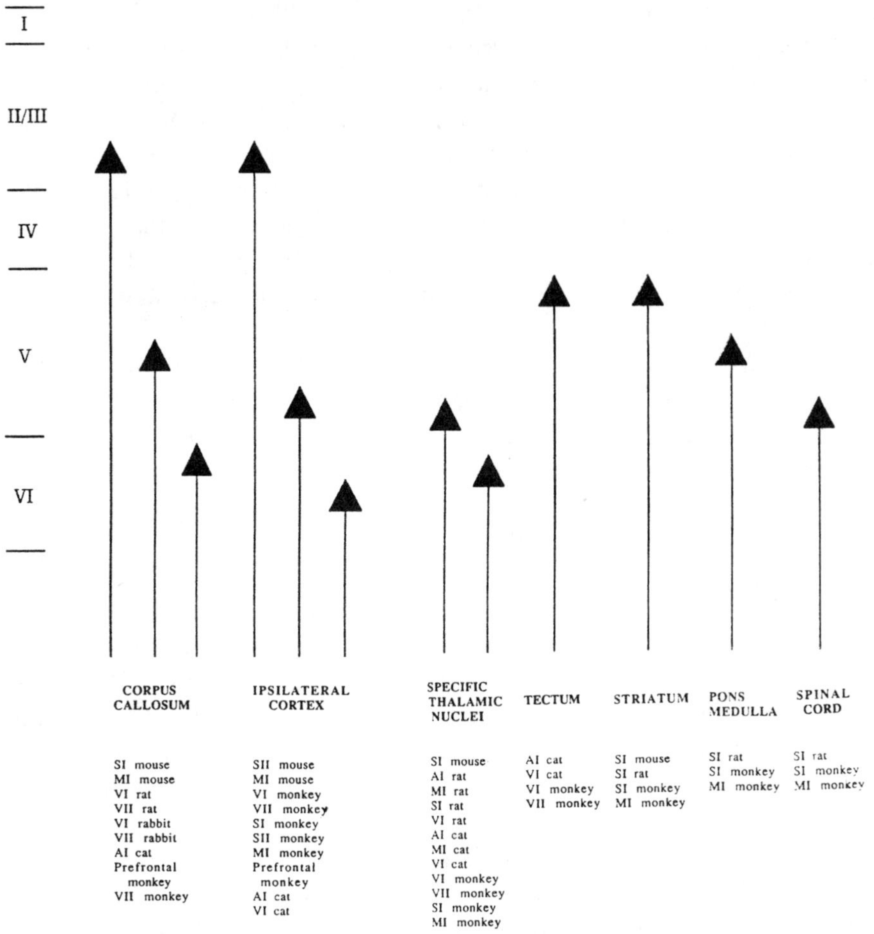

Figure 1-1. Laminar distribution of pyramidal neurons (black triangles), listed according to their projection target areas. Listed under each of the latter are the cortical area and species in which the projections shown have been reported. Reprinted with permission of Birkhäuser Boston, from White (1989): *Cortical Circuits: Synaptic Organization of the Cerebral Cortex*, pp. 26–27.

In contrast, layers II and III appear to mediate short- and longer-range intracortical communication, respectively.

We may therefore say that columns in most areas mimic the basic organization of those in area 4, even though many of them lie at some distance from the final motor pathways. This mode of organization perhaps reflects the priority given throughout evolution to motor functions, which is to say, to outputs.

Hence, in vertebrates generally, the vestibular system is the first to come on-stream developmentally (Fair, 1987). And in the ontogenesis of the neocortex, e.g., in man, layers VI and I are laid down first—layer I representing a rostral extension of the ancient reticular formation which, as Bishop (1958) noted, extends the length of the neuraxis. Layer V differentiates from VI; the cells forming layers II–IV then migrate through V–VI to their final positions, layer II being, in this respect, the "youngest" (Sarnat, 1987). Even the ontogeny of single units reflects this order of priorities, development of the axon coming first, with differentiation of the dendrites occurring much more gradually (Sarnat, loc. cit.).

As suggested by this author (Fair, 1965; 1988), the precedence given to motor functions (or action) over sensory processing (or perception and cognition) may, during evolution of the vertebrate CNS, have made memory a function of output arrays wherever these happen to be located. That is, even in sensory systems, it is the most frequent *outputs* that tend to be retained. In other words, memory functions may be, if not specifically, at least most critically a characteristic of magnocellular relay assemblies such as those of layers V and lower III in neocortex. Recently reported data supporting that conclusion are reviewed in parts 1 and 2 of chapter 5, below.

When Pons et al. conclude that the transcortical relay of information from prime receiving cortex to parasensory and association areas mainly involves large pyramidal arrays in lower III that project mainly to lower III elsewhere, they may therefore not simply be defining a sensory throughput system. These III-to-III connections may also represent links or successive stations in a cortical memory network. The grounds for this conclusion, and for my supposition that this sort of relay results in stepwise increases in the generality of the corresponding memories, are given in the chapter that follows.

Apropos of the foregoing, a report by Iriki et al. (1989) indicates that in the cat tetanic stimulation of postcentral (somatosensory) cortex results in long-term potentiation (LTP) of units of precentral (motor) cortex. Using electrolytic lesions as histological markers, they found that "associative LTP was induced only in cells in layer III."

Noting that layer III cells are "noncorticofugal," they suggest that these may then figure in motor learning by modulating the output "of a small group of cortical efferents," i.e., chiefly units of V–VI. In support of this conclusion, they cite studies showing that interruption of connections between somatosensory and motor cortex retards motor learning in cats, without affecting motor skills acquired preoperatively.

They propose that associative LTP may figure in motor learning as a corrective feedback mechanism by means of which, during practice, output from somatosensory to motor cortex progressively refines the resulting performance, essentially by modification of the motor memories in which performance originates. At criterion, such feedback may no longer be necessary, and performance will then largely be triggered from ventralis lateralis.

The fact that these workers were unable to obtain LTP by tetanic stimulation in the thalamus might be due to technical difficulties; if not, however, it carries an interesting implication—that transcortical (i.e., serial) processing may be chiefly involved in memory formation. Once the memories formed this way result, as in the preceding illustration, in adequate performance, the need for transcortical transmission diminishes. Thus, in the visual system, parallel processing of inputs via the pulvinar may not figure significantly in the *formation* of visual memories but may act as an efficient mechanism for their activation, once formed. (See chapter 5, part 3, below.)

And if, in general, new learning depends upon transcortical processing, whereas habitual sensory or motor responses can be mediated more directly (hence more focally) from the thalamus, it follows that in the early stages of conditioning we would observe desynchronization in the EEG, whereas, postcriterion, the EEG response would become highly focal. This is indeed the sequence usually observed, and one that appears to be consistent with the single unit studies of Gabriel et al. (1980). These showed that during training to criterion in an operant paradigm, focal activity chiefly involved layers V–VI in rabbit cingulate cortex. When the response had reached the postcriterion or habitual stage, the principal focus appeared to be in the anterior thalamic nuclei.

The relation of LTP to long-term learning involving the hippocampus is taken up in chapters 4 and 5 below.

Chapter 2

The Role of Magnocellular Bands in Generalization or Categorization of Sense Data

Abstract. Evidence that magnocellular pyramidal arrays have generalizing functions; and that the memories formed in this way represent the most probable forms taken by the organism's day-to-day sensory experience. How successive generalization results in a noise-resistant retrieval system, facilitating recognition of a given auditory or visual input as the "same" under a variety of transforms. How essentially the same process leads, in parietal cortex, to stepwise increases in the generality, or finally in the abstractness, of spatial representation. Data illustrating these points are cited from the psychological and clinical literature.

The typical structure shown by pyramids of lower III or V (see, e.g., Lorente de Nó, 1951, p. 296; Jones, 1981, p. 206; Lund, 1981, p. 107) suggests that they act as "integrators," in the sense that their axonal output is a resultant or weighted algebraic sum of inputs to their somata and apical and basilar dendrites. Because of their relatively great size in many areas of generalized eulaminate cortex (Bailey and von Bonin, 1951), the output of such units is likely to reflect convergence of inputs from a relatively large number of sources, both immediate and distant.

Estimates, made by electron microscopists, of the average number of synapses found on cortical neurons are very large—on the order of $1\text{–}3.8 \times 10^4$ (Colonnier, 1981). In the case, for instance, of large pyramidal relay units of III and V in primary visual cortex, these synaptic inputs appear to be systematically organized, as described by Lund (1981). This feature, while making for some degree of response-specificity, would not necessarily rule out the possibility that N different combinations of synaptic input ($N \gg 1$) might result in the same all-or-none output.

A way out of that difficulty was suggested by Blakemore's proposal (1974) that units in visual cortex or elsewhere, might tend over time to respond differentially to inputs they had "seen" most often. Such probabilistic response-biasing would mean that at an early developmental stage, units, e.g., in area 17, which had, *ab initio*, shown diffuse responses would begin to become more sharply selective, to the point that in the mature animal, the system would appear to be "hard-wired."

Recent evidence shows that in fact it is not—that it acts as Blakemore has suggested, and therefore retains some of that plasticity even at maturity. In two studies, Kaas et al. (1981; 1990) demonstrated that partial deafferentation of the hand area in somatosensory cortex in adult macaques, or limited deafferentation of 17–18 by laser lesioning of spots, 5–10° in diameter, just above the area centralis in the retinae of adult cats, produced cortical areas that were at first "silent" to stimulation, but in which, over a period of weeks, neurons developed new, atypical receptive fields.

(There were limits to this type of functional recovery. If, for example, in cats, retinal lesions were 10° to 15° in diameter, neurons at the edges of the "deprived" cortical areas might develop new retinal fields, while those towards the center remained unresponsive. Likewise, in the macaque, "removing inputs from more than half of the hand produces a larger zone of deprived cortex where complete reactivation does not occur.)

A similar phenomenon, produced in quite a different way—as a side effect of visual deprivation—was reported by Creutzfeldt and Heggelund. (Their experiment is analyzed in detail in my recent book; Fair, 1988, pp. 31–35. See also Bear et al., 1989, who present what amounts to a mathematical model of Blakemore's principle.).

The point is important, since it implies that, beginning with the prime receiving areas, a kind of memory mechanism is at work, such that any assembly initially has a certain range of inputs available to it, out of which, over time, it selects those having the highest *de facto* probability.

It is essentially this change that we may see taking place as an organism makes the developmental transition from diffuse visual responses of the "aspecific" or "immature" type (Cooper, 1981) to those of the seemingly "hard-wired" type reported by Hubel and Wiesel (1962; 1963; 1964) or many others since.

While the probabilistic biasing of single unit output may be far more complex (see for instance, Poggio and Torre, 1981) than the foregoing description might suggest, particularly in light of data as to neuropeptide modulation, the evidence just reviewed, together with some reported by Pribram (1989) suggests that Blakemore may be right. Such biasing may then also automatically give rise to generalizing or categorizing functions, as a result of the convergence of parvocellular input upon larger units such as the pyramids in magnocellular bands of V and lower III.

For example, Gilbert and Wiesel (1981) report that, in striate cortical columns, "the predominant intracortical projection from layer IV appears to be to layer II + III, which would then represent the second level of cortical processing. Layer II + III contains complex cells almost exclusively.... The layer V complex cells may form their fields from the concatenation of the fields of the superficial complex cells, as suggested by the extensive projection from the layer II + III cells into layer V."

While in area 17, both III and V are relatively lacking in the type of large pyramidal arrays seen in association areas, these findings imply that the larger units in these layers may "concatenate" their more frequently "seen" inputs, preparatory to relay, as described by Pons et al.—i.e., chiefly from lower III to lower III elsewhere.

That the larger cells in III and V may, because of their convergence of inputs, have generalizing functions, is suggested by Gilbert and Wiesel's description of special complex cells, which respond as well to slits 1/8° as to one the length of a receptive field (~3°). They are reportedly found at the bottom of layer III and in V, and have a generalizing function in that they respond to slits having a range of lengths but a common orientation.

In other words, what Gilbert and Wiesel may be describing is the start of a several-stage process in which highly fractionated visual input data (cf. Livingstone and Hubel, 1988) are sorted into clusters of most frequently "seen" invariants. It is essentially this process, repeated intramodally and finally (in areas such as 7b or 39) cross-modally, that may result in the type of memory system proposed here—one whose contents show a stepwise loss of detail and a corresponding increase in generality, i.e., in the range of particulars they "stand for."

In later stages of the visual relay—for instance in MT (the middle temporal area) in the owl monkey—receptive fields tend to become much larger (Allman, 1987). And whereas units (e.g., of 17 or V1), having what Allman calls "classical receptive fields," respond to specific directions of movement, motion-sensitive units in MT have been shown to respond to the amount and direction of relative movement, for instance of a bar, surrounded by a stationary or moving pattern of dots.

Such units, Allman notes, show responses that have become more "global," meaning context-dependent. The context is provided by a pooling of data, such that a given MT neuron receives input from N "classical" direction-sensitive units ($N \gg 1$) representing a common orientation.

A similar, but more complex and variable, pooling of data appears to occur in inferotemporal cortex of the macaque (Gross et al., 1974). According to their report, "Every receptive field included the center of gaze, the fovea." Otherwise, the receptive fields were highly variable, with over half extending well into both visual fields, about one third being in the contralateral, and the remainder in the ipsilateral, hemifields. They were very large, "usually more than 10° × 10°." For some of these neurons, the "best stimulus appeared to be a monkey hand, the shadow of a hemostat, a bottle brush or a specific curvature." The researchers concluded that some of these responses, for instance "to a 1°× 5° red slit, oriented at 45° within its receptive field" might have been part of a larger pattern, equivalent to a "perhaps meaningful stimulus we never used."

In other words, the trigger stimulus in this case may have corresponded to one piece of a visual memory. Other pieces, associatively related to it—i.e., which would tend to be concurrently activated as parts of the same memory—might include units representing those areas of the visual field in which the complete input was most likely to arise. An example might be the upper visual hemifields in the case of an edible red fruit the animal most often found above its head.

This amounts to saying that, within certain constraints, the size of visual receptive fields, and related properties such as color responsiveness of units in this cortex, become organized probabilistically. The potential for a variety of associative combinations is there; experience determines which combinations actually take shape. And in general, the combinations will tend to represent adaptively important features of the animal's sensory experience in the form of type-objects or categorical memories (cf. Wilson and DeBauche, 1981).

In contrast to the response characteristics of units in primary visual cortex, which in a given species are relatively predictable from one animal to the next, those of units of inferotemporal cortex may be much less so. The units sampled by Gross et al. (1974) suggest not a topographic ordering but a miscellany, reflecting the miscellaneity of the animal's day-to-day visual experience.

The response properties Allman reports for units of MT suggest a less flexible part of the visual system whose motion detection functions may develop early and thereafter form part of a fixed repertoire. It would still be of interest to know the extent to which these are learned and where, in the developmental timetable, that learning is apt to be complete. Would owl monkeys raised under conditions that minimized their exposure to conditions requiring them to estimate the relative motion of objects around them, show significantly fewer units in MT having the properties reported by Allman?

For cytoarchitectonic reasons discussed in the next chapter, generalization of data, via the III-to-III system described by Pons et al. (1987), may become maximal in final stages of the relay. These are the stages involving entorhinalis, the FD, hippocampus, and return pathways to the cortex. Not only is this loop apparently essential to the formation of long-term "declarative" memories. It may also serve to establish a set of key descriptors that, as part of LTM in the cortex, permit "retrieval" or activation by playback of the less generalized data to which they correspond.

The way the system works is crudely illustrated in Figure 2-1. I must emphasize that the figure is intended only to suggest the type of generalization or data pooling mediated by large-celled arrays upon which the output of smaller-celled arrays converges. This *kind* of operation is then organized modularly, e.g., as just shown in the case of relative motion detection units in area MT of the owl monkey, or as is shown below for the parietal representation of space.

The numbers in the top tier, and the three tiers of boxes below them represent assemblies (cortical columns or groups of them). The connecting lines represent the layer-III-to-III relay system of Pons et al., their convergences representing the most probable or commonly associated features of a given sensory input.

If we imagine the diagram as extending out of the plane of the page to include assemblies $1'$, $2'$, $3' \ldots n'$ (some of which cross-connect with those shown), what 15, at the D level, specifies at the A to C levels may also extend out of the plane of the page; and some of those specifications will overlap with those made by assemblies at the D level elsewhere.

If we run this system backwards, as a retrieval network, it is then the particular combination of most general features, represented at the D level, that acts, by cross-convergence at higher levels, to activate the most probable combination of less generalized data to which they correspond. In other words it acts as a filing system. In the input mode, fractionated sense data are put through a succession of comparators—first for best fit of small components, then for best fit of combinations of these, combinations of combinations, etc.

In each case, the criteria for best fit will, in the mature organism, have been established probabilistically. At the far end of the comparison process, either a set of existing final descriptors will be activated or new ones may begin taking shape. That is, in this mode, activity in the network converges.

It is via this system, in the input mode, that we recognize four-legged more or less hairy land animals that bark as probably dogs, even if, for neurological reasons we may not be able to find their generic name. In the output mode, the same system enables us to start with the word or idea "dog" and associatively call up the various features that, in our experience, define one. That is, in this mode, activity in the network diverges.

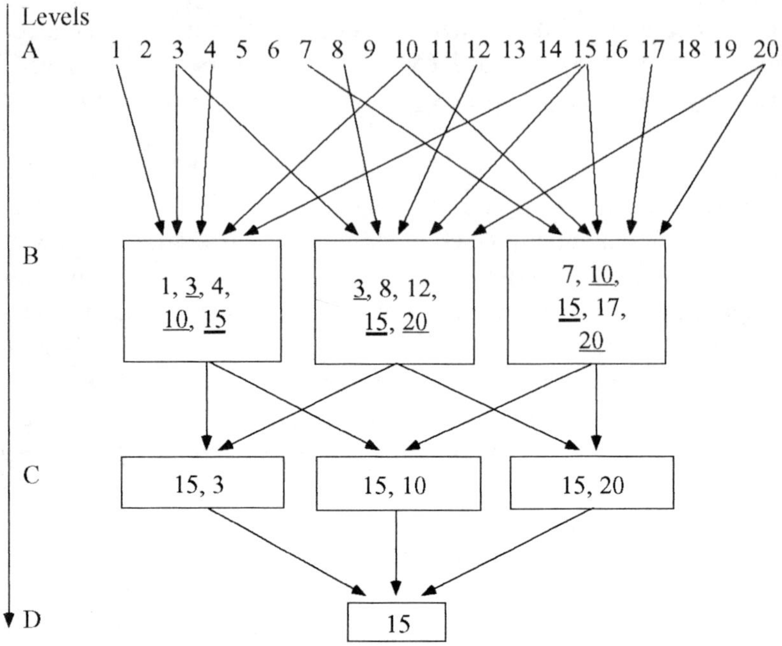

Figure 2-1. The numbers in A represent coded features of sensory data relayed to assemblies at level B. The feature having the highest probability of being received concurrently with others at that level is item 15 (heavy underlining). The next most probable features are each represented in two out of the three assemblies at level B (underlined items 3, 10, and 20). Features of still lower probability are items 1, 4, 8, 12, 7, and 17. They correspond to features least frequently encountered in the sensory contexts represented in these assemblies.

In the assembly shown at level D, only the most probable common feature (15) is represented. If we imagine D as extending out of the plane of the page and laterally in either direction in it, items in D correspond to a set of descriptors, such that if the system is run backwards, the combination of descriptors will act, as in filing-systems, to retrieve certain data sets. For example, if 15 in D here corresponds to the feature four-leggedness as in various animals, it might also, if activated concurrently with other items at the D level, correspond to four-leggedness as in inanimate objects such as tables.

It is this kind of invariance sorting, occurring as a result of exposure to experience and conserved as probabilistic response biases in the several-stage interareal relay system, that gives rise to generic memory. This is the system by which we come automatically to recognize the categories into which the beings, objects, or events of everyday life fall. How it leads to our effortless use or understanding of figures of speech, and provides us with some of the raw materials of abstract thinking, is described in Chapter 8.

For a more sophisticated geometrized model, showing how a several-layered network can recognize a visual input as the "same" under transforms at the primary (retinal) level, see Poggio and Girosi, 1990.

There is still another point to note. The interareal relay system that permits auditory or visual recognition of inputs by category may or may not involve formation of long-term memories representing, say, a specific dog. One can have known what dogs are without ever having come to remember, and hence recognize, a particular dog. This final tier in the file need not be filled for the file to work.

If, as I have proposed, units subject to probabilistic biasing on Blakemore's principle tend to be concentrated in large-celled output arrays; and if, as Pons et al. suggest, such arrays are a principal means of interareal relay, the result will be the type of memory system just outlined. It is one that, given sufficient exposure to experience, leads almost automatically to generic recognition of a wide variety of sensory inputs.

This view appears to be supported by data from studies of apparent phoneme recognition in human infants 1 to 12 months in age. Eimas (1985) suggests that, since we are able to recognize a given sequence of phonemes under a very wide range of transforms, "we perceive speech categorically." We do so because, day-to-day, the relay of input in auditory cortex generalizes from these data, extracting invariances as to timing, amplitude, and pitch of most commonly heard sounds.

When, by further relay, this process is extended to include commonly heard sequences or patterns of sound, the corresponding auditory memories will become more generalized, in the sense that any given recalled pattern will then correspond to a certain *range*, e.g., of time intervals, pitches, or amplitudes in its components. It is essentially by successive relays of this type, that we become able, within certain limits of distortion, outright mispronunciation or extraneous noise, to identify heard words correctly. In effect, this is a memory system that, by generalizing over time from highly variable or fuzzy data, adapts to the fact that any spoken word is, auditorily, a fuzzy class. Though somewhat error-prone, the system works.

(See Figure 2-2. The scheme suggests the contribution of lateral, but not of fed back inhibition to if-then processing. For a computerized model showing how both types of inhibition result in the kind of probabilistic sorting described here, see Ambros-Ingerson et al., 1990.)

Figure 2-3, adapted from Eimas, shows what he calls "the waning of unused perceptual powers." Whereas at 6–8 months, children from English-speaking families responded to phonemes (uncommon in English) derived from Hindi or Salish (an American Indian language), these responses progressively fell off when they were tested at 8–10 and 10–12 months of age. In children born to families speaking these languages, responses at 11–12 months were stable at the 100% level.

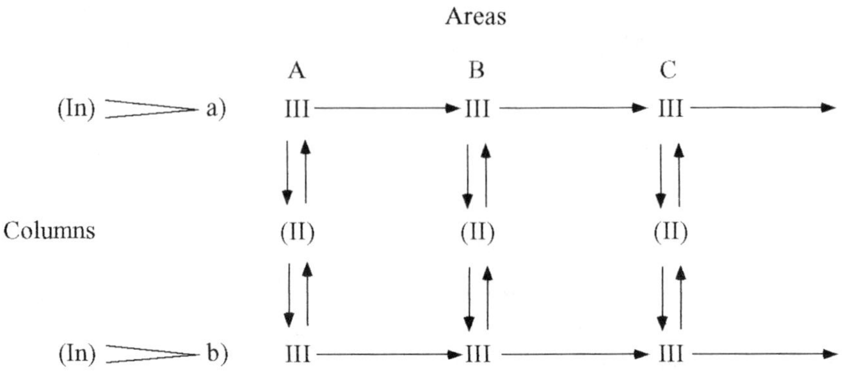

Figure 2-2. In the schematic, a and b represent small adjacent assemblies or columns in sensory cortex. The arrows connecting III represent the interareal relay system. A, B, and C represent stages in the relay. The vertical arrows represent the shorter intra-areal projections arising in cortical layer II. It is proposed that these act to establish associations between memory items within a given area, by way of cross-linkages equivalent to probabilistically derived concurrence (if-then) relations, including their inhibitory negatives.

Because data relayed via magnocellular lower III are successively stripped of detail, leaving successively larger regularities (patterns of patterns of patterns, etc.), a system such as the auditory is capable of identifying, via these higher-order invariances, the words uttered by a given speaker, even when grossly mispronounced. The fact that we are used to hearing language pronounced in a variety of ways structures our verbal memory system accordingly, enabling us to hear a broad range of phonemic variants as the "same."

(The children from English-speaking families showed a small, but consistent tendency to respond more frequently to Hindi, an Indo-European language, than to Salish, perhaps reflecting the fact that the phonemes from Hindi were slightly more like those of English.)

Eimas suggests that auditory categorization "occurs because a child is born with perceptual mechanisms that are tuned to the properties of speech," a conclusion he thinks supports Chomsky's notion that "inborn knowledge and capacities underlie the use of language." In fact, Eimas's data support the opposite conclusion. The categorizing functions he describes, far from being peculiar to speech sounds, or to human beings, may be a general property, e.g., of primate association cortex proximal to prime receiving areas such as 17–18 or 41–42.

If so, the experiments Eimas reports should be repeatable with infant chimpanzees or macaques. The model of memory organization presented here would predict that exposure of these animals to phonemic combinations derived from the language of their keepers should result in responses

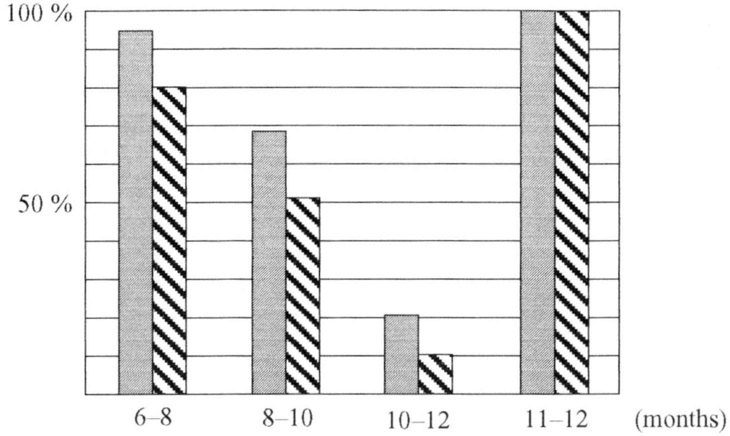

Figure 2-3. The histogram shows the results of a study in which infants from English-speaking households were tested for their responses to "consonantal contrasts" in two unfamiliar languages, Hindi (left member of the paired bars) and Salish, an American Indian language. Tests were given, and subsequently repeated, at ages 6–8, 8–10, and 10–12 months. At right are shown the responses, at age 11–12 months, of infants raised in households in which either Hindi or Salish were spoken.

In the children from English-speaking families, the sharp decline in per cent of responses to Hindi and Salish between ages 6 and 12 months suggests that their initial responses to those language sounds represented a form of incipient learning that was then progressively crowded out by the more commonly heard sounds of English. It is interesting but probably not statistically significant, that their responses to Hindi, an Indo-European language, were consistently greater than those to Salish.

Reprinted with permission of Scientific American, from Eimas, PD (1985): The Perception of Speech in Early Infancy. *Scientific American* 252:46–52.

closely resembling those of human infants. This would particularly be the case if, pre-experimentally, the animals had been spoken to often, and had been rewarded upon hearing certain words. If they then ceased to be spoken to and were tested again several months later, they might also show the "waning of unused perceptual powers" described in the experiments above.

Should these experiments turn out as predicted we might have to conclude that the auditory categorization of phonemes is neither unique to man, nor limited to language sounds. (We categorize other sounds, and sights as well.) Categorization occurs simply as a consequence of the way the posterior sensory relay systems work.

If there are "modules" responsible for the appearance of language in man, they are probably the supramarginal and angular gyri and Wernicke's area—40, 39, and 22 in posterior cortex—and Broca's area, usually defined as 44 and 45 frontally. (See Fig. 2-4.) Mesulam (1985) notes that Brodmann's study of the monkey brain does not include any areas numbered 39 and 40, and Crosby et al. (1962, p. 453) state that these "have been identified only in the highest primates."

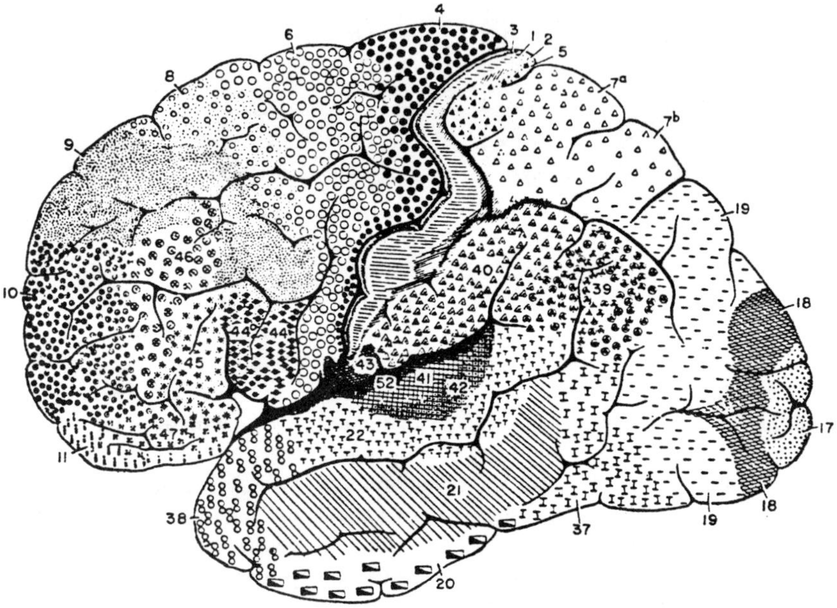

Figure 2-4. Cortical cytoarchitectonic fields in lateral view, as defined and numbered by Brodmann.

The importance of these areas lies in the fact that they represent convergence of input in three modalities. Recall also that in strips rostral and caudal to the central sulcus the body has motor and sensory representation, respectively, and that that "homunculus" is contralateral to the real one, and also upside down. The leg, foot, toes, and genitals are represented dorsally and medially; the arm, hand, and fingers laterally; and ventral to these, the eye, nose, face, lips, teeth, jaw, tongue, and (ventralmost) the abdomen.

These arrangements make functional sense, not only as suggesting a basis for foot fetishism, but as indicating how verbal functions arise by way of loops connecting area 22 (involving verbal memories) with 39 and 40. The arcuate fasciculus (not shown) then links these areas with 44–45, which are homologously placed association areas on the motor side. (The way these frontal and posterior systems work together is discussed in Chapter 6.)

The relevant point here is the way in which generalization from primary body-sense data may occur in this posterior cortical system. Consider the parietal representation of space. In the squirrel monkey, vestibular input reaches area 3a by way of the posteroventral nuclei (Guldin et al., 1989). Farther away, in parietal 7b, Mountcastle (1986) has reported units in the macaque responsive to both visual and somatosensory stimulation. In parietal 7, Guldin et al. found only a few units showing vestibular responses, and none in primary visual areas.

Functionally, what this shift from vestibular to visual association means is that whereas the representation of space in primary somatosensory cortex is directly related to (changes in) position of the body, in parietal association cortex it expands so as to relate body-sense data to the larger visual world in which the body operates. Hence unilateral damage to this cortex in humans can result in "neglect" of the contralateral "extrapersonal space." (Mesulam (1985) notes that damage to the superior parietal lobule, rather than to the inferior, "may be more closely associated with neglect behavior.")

In the inferior parietal lobule, especially the angular gyrus, which Geschwind (1974) called the "association cortex of the association cortexes," spatial representation may then undergo another transformation. That is, in cross-modal memories at this stage in the relay, the associational component representing space may become more abstract, and also more associatively linked to the representation of temporal sequences

In written sentences, word order is spatial, but because, like spoken sentences, they are sequential in time, verbal "space" tends to be unidirectional. That is, it is difficult but not impossible to treat this space as traversable in the other direction—to speak or write sentences or play a piece of music backwards.

This amounts to saying that the spatial component in associative memories, e.g., of words, may derive from second order visuospatial memories formed in areas such as parietal 7b. The latter represent space in general, in the same way that the generic visual memory "dog" represents dogs in general.

By further relay, e.g., to distal association cortex such as the angular gyrus, these visuospatial data then provide an abstract coordinate system in which we arrange a variety nonspatial items. Hence we speak of prices or musical tones as "high" or "low." Many metaphorical usages of this type can be traced to their bodily origins.

High pitched vocal tones, like high places in the environment, require more physical effort to get to than low ones. "High" by further abstraction becomes a state of eminence in life often requiring exceptional effort to reach, and we often say we "look up to" people who reach it. Figures of speech of this type are, I believe, a direct result of the way in which our memory systems—in this case, of the parietal lobe—are built up by the several-stage processing of primary sense data.

Let me be clear about what is meant by the terms "generalized" and "abstract" as used here. A generalized or generic memory is one representing certain statistically salient features of some class of beings, events or objects. It is a crude approximation of the memory that would emerge if all possible members of the class could be sampled.

Abstract memories derive from those of the generic type essentially by the same statistical process, but one that is limited to certain of their features. That is, it selects (abstracts) from the salient features represented in generic memories, a subset representing features most salient (probable) over a certain range of associatively established contexts. That is, it is a biased higher-order sorting of already sorted data.

At the rostral end, then, of the cortical relay in the parietal lobe, we have the body-centered space representation mediated by the vestibular and proprioceptive systems. More posteriorly (e.g., in 7b), visual and somatosensory inputs combine to represent extrapersonal space. And finally, in areas such as the angular gyrus, we find the abstract time-linked space representation just described.

However, the system as a whole is a linked one, which means that if we are to remain aware of the (real external) space represented in our right and left visual hemifields, relays from the visual system to corresponding areas of parietal cortex need to be intact.

When those connections are cut unilaterally, for instance as a result of ischemic damage following stroke, the result may be a patient who, although otherwise apparently rational, behaves as though half of the (contralateral) space around him and half of his own body no longer exist. A particularly interesting case, involving unilateral neglect of *abstract* hemispace, is reported by Caramazza and Hillis (1990).

The patient, a 77-year-old left-handed woman, had had a stroke involving "a large area of the left parietal white matter and left anterior basal ganglia." Her spoken language, eye movements, and visual fields were all normal. She had "severe difficulty in processing the right side of objects and words."

She could correctly name all the letters in a written word and still make errors in the right hand part when reading it. The errors were dependent upon position—that is, upon an abstract space consisting simply of "first half, last half." Whatever the length of the word, she split it into approximate halves, being apparently unable to see the last half.

For example, she read "contrastiveness" as "contrasting"; but when given "contrast" alone, shifted the error to produce "continue" or "control."

She made the same kinds of errors in written, oral or backward spelling, and in reading mirror-reversed or vertically arranged words. When asked to bisect a straight line, she shifted the midpoint to the left, even though she began by correctly marking the ends of the line.

In this patient, abstract space was divided into hemifields as though, in the course of her maturation, it had derived by parietal relay from the hemifields of literal bodily space. The resulting perceptual fields were abstract in the sense that they telescoped the usual right–left, up–down, backward–forward distinctions into a single axis. (Her reading of words was unaffected by their being presented in column or as mirror images.) She could find the ends of the line, a task that could be done in single tries, each equivalent to "first half." However, any sequential perceptual task tended to stop midway for want of a "space" to contain the second half. Left parietal damage appears to have selectively damaged this contralateral derivative of literal bodily space.

The way in which these several representations of space arise in the parietal lobe is thus analogous to the type of memory formation underlying the "auditory categorization" of phonemes reported by Eimas (1985). It differs in being a cross-modal form of generalization by means of which vision expands the sense of space due to balance reflexes and proprioception to include the extrapersonal space of the distance receptors.

In contrast to Caramazza and Hillis's patient, who exhibited a true, if unusual, case of unilateral neglect, patients with Balint's syndrome have a disorder that Damasio (1985) describes as "an acquired disturbance of the ability to perceive the visual field as a whole." Instead, they show an "unpredictable perception and recognition of only parts of it... accompanied by an impairment of target pointing under visual guidance... and an inability to shift gaze at will toward new visual stimuli."

The disorder, he reports, most often occurs as a result of bilateral infarctions "in the border-zone (watershed) between the anterior and posterior artery territories" and may or may not be accompanied by inferior visual field defects, depending on the degree of involvement of visual area 19. Even if such defects are not present, the patients find they can focus on "only a small fraction of the panorama, outside of which vision is hazy." Moreover this fraction "is not stable and moves erratically from quadrant to quadrant." (Damasio, loc. cit., p. 276.)

The cortex principally affected is parietal, e.g., of 7b, a region homologous to the one in which Mountcastle (1986) found bimodal (visual and somatosensory) units in the macaque. Damasio adds that besides medial or dorsal parietal cortex, parts of 19 or 39 (the angular gyrus) may also be involved. Balint's syndrome is thus a disturbance of "extrapersonal" space, apparently resulting from lesions that bilaterally cut connections between inputs representing external visual space and a coordinate system derived from (literal, bodily) space as represented in 3-1-2.

These patients contrast sharply with those having "bilateral lesions confined to the supracalcarine cortex, thus sparing the adjoining parietal cortex" (see Fig. 3-1, below). In these cases, Damasio reports, the patients "claim not to see objects they can point to accurately." The reason is that the parietal representation of extrapersonal space has been spared, and visual information can still reach that cortex, e.g., via the thalamic association nuclei, or by way of the temporo-parietal projections mentioned in the next section.

The latter projections evidently figure in the visual perception of motion, and patients with Balint's syndrome often complain "that moving objects are especially difficult to perceive." By contrast, their perception of literal bodily space (involving areas closer to 3-1-2) is often relatively intact. They are able to point accurately to parts of their own bodies or to their clothes. They can also locate sounds, evidently because cortex handling audio-somatosensory associations is not involved.

The peculiar perceptual problems accompanying Balint's syndrome, even in patients without apparent defects in vision per se, illustrate the principle outlined above, according to which, by way of the relay from primary somatosensory cortex to areas such as 7b, data relating to the body's position in gravitational space become systematically associated with visual inputs so as to represent extrapersonal space—i.e., space in general. There are even gradations in the process, such that "peripersonal" space (space within reach) is differentiated from "far" (or true extrapersonal) space (Halligan and Marshall, 1991; see Appendix IV for details).

The associative connection between visual and body-related space is, however, so close that when visual inputs cannot access their cross-modal referents in posterior parietal cortex, visual perception loses its compass. Shut off from that basic coordinate system, the eyes do not know where to look—the condition Balint himself called "psychic gaze paralysis" (Damasio, loc. cit., p. 277).

It is perhaps only adaptive that that should be the case. Animals whose perception of extrapersonal space did not remain tightly linked to data representing their own position in it might be prone to walking off cliffs.

In the chapter that follows, I will take up the processes leading to LTM formation and thereby to specific recognition.

Chapter 3

The Hippocampal System and LTM

Abstract. The previous sections concern a generic memory system that takes shape "forward"—that is, by successive relay from prime receiving to parasensory and association areas. This Section concerns the long-term memory system that mediates specific recognition and depends upon pathways that reciprocally connect association cortex with the hippocampus. To the extent that their formation depends upon return fiber systems or playback from the hippocampus, long-term memories are laid down "backward." Consolidation is thus a recursive process, and in the absence of the hippocampus does not occur.

The structure of such memories is counterintuitive in three respects: (1) they consist of elements already existing in the generic memory system, most of which are subject to further probabilistic change; (2) only part of a given long-term memory is subject to fixation; and (3) the parts fixated consist of elements represented at the "far" end of the generic memory system—elements that, being maximally generalized, are in themselves least liable to change over time. The psychological consequences of this arrangement are explored in later sections.

In the hypothesis proposed here, a critical assumption is that "circular reinforcement" over loops connecting the hippocampal system to parts of neocortex is essential to the formation of long-term memories of the kind that mediate specific recognition.

The topics covered in this section will include:

(1) biochemical features of the hippocampus that reflect its importance as a monitor of systemic and basal central nervous states

(2) the type of information processing implied by the structure of the pathways that reciprocally connect it with the cortex; and

(3) clinical evidence that appears to show that memories mediating specific as opposed to generic recognition have the differential cortical distribution predicted under the present hypothesis.

It is generally thought that what we most lastingly recall is apt to be related, in some direct or indirect way, to events, persons, etc., that are or were "motivationally" important to us. In other words, they generated an emotional or intentional climate sufficiently intense to cause them to be selected for incorporation into long-term (specific) recall, or LTM.

A considerable body of evidence suggests that the dentate gyrus (FD) and hippocampus play a critical role in that selection process. Hence bilateral elimination of the hippocampus, as a result of surgery or ischemic damage, has been found to preclude further additions to LTM in man (Squire, 1986, 1988). Sass et al. (1991) report that in epileptic patients, although cell losses occur in all three hippocampal subfields as well as in the hilus and dentate gyrus (FD), shorter-term memory deficits in these patients correlated specifically with pyramidal cell losses in the CA3 field.

The hippocampus is an ancient structure whose primordia predate the appearance of mammalian neocortex. It is also one that, like the basolateral nuclei of the amygdala, shows evidence of being phylogenetically "progressive" in higher vertebrates (Angevine, 1965; Sarnat and Netsky, 1981), as does the FD or dentate gyrus (Bayer et al., 1982).

In functional terms, what this means is that, while retaining its responsiveness to neuronal and/or neurochemical inputs representing basal body states, the hippocampus has also developed neocortical connections that extend its influence rostralward, making it critical, in a double sense, to the selection process just mentioned. Not only does it "decide" what concurrent sense-data may warrant indefinite retention at the cortical level; it is also part of the circuitry, and may provide the neurochemical means (e.g., the neurotrophic factors discussed in Section 4) that make such retention possible.

Consistent with its function as a kind of global monitoring system, the hippocampus has important brainstem connections via the mammillary bodies (Fig. 7-1 below), and contains both Type 1 and Type 2 glucocorticoid receptors. The Type 1 receptor has limited forebrain distribution and apparently includes subtypes that bind the mineralocorticoid aldosterone and its precursor-in-synthesis, corticosterone (De Kloet et al., 1987). The Type 2 glucocorticoid receptor has quite wide distribution in the cortex, diencephalon and limbic system (McEwen et al., 1987). It is found on catecholaminergic and serotonergic units, and on neurons showing CCK, β-endorphin, enkephalin, somatostatin, or CRF immunoreactivity, to name a few (cf. Fuxe et al., 1987). Table 3-1 shows, in descending order, the respective densities of Type 1 and Type 2 glucocorticoid receptors at several limbic, diencephalic and brainstem sites in the rat (De Kloet et al., 1987).

The high to maximal density of Type 1 receptors in the hippocampal system is of particular interest, since the affinity of this receptor to glucocorticoids is reportedly 6 to 10 times higher than that of the Type 2 (De Kloet et al. give the lower estimate; McEwen et al., 1987, the higher.)

The Type 1 receptor (originally reported by McEwen et al., 1968) may figure quite directly in memory formation arising out of stress. McEwen et al. (1987) found that, in adrenalectomized rats, corticosterone raises

Table 3-1.

Subcortical glucocorticoid receptors, by locus, in descending order of density:

Type 1	Type 2
CA1/subiculum	Lateral septum
Dentate gyrus	Dentate gyrus
CA3	N. solitarius
N. solitarius	Central amygdala
Lateral septum	Locus coeruleus
Locus coeruleus	CA3
PVN	Supraoptic n.
Central amygdala	PVN
Supraoptic n. (0)	CA1/subiculum

PVN: paraventricular nucleus; N. solitarius: nucleus of the solitary tract.
Source: De Kloet et al., 1987.

synapsin levels in the hippocampus but not elsewhere in the brain—a result, they note, that has been obtained in other laboratories.

Synapsins are "presynaptic phosphoproteins which may be involved in release of synaptic vesicles. Increased synapsin levels... may indicate increased numbers of synaptic terminals or increased synaptic terminal size." Synapsins are thus likely to be one of the components of long-term potentiation (Chapter 4), and so to figure in the induction of LTM by way of hippocampal output to the cortex.

In connection with the increased synapsin synthesis evidently mediated via the Type 1 glucocorticoid receptor, it is well to recall a principle stated by Stryer; that the "primary effect of steroid hormones such as estradiol, progesterone and cortisone is on *gene expression* rather than on enzyme activities or transport processes" (Stryer, 1981; italics original). As will be seen in later sections here, this principle may be reflected in the fact (1) that, to occur, LTM formation seems often to require some minimum of stress or accompanying central activation; and (2) that the process itself may involve activation of immediate early genes (IEGs) and their mRNAs.

De Kloet et al. (1987) note that "the highest Type 2 receptor density was detected in neurons known to have an important function in the stress response." In the septum and hippocampus, they "seem to be co-localized with Type 1." There is neuroanatomic and some physiological evidence that hypothalamic projections from the hippocampus or from the subicular area may provide negative feedback acting to limit rises in ACTH output. (McEwen at al., 1987; see also Mason, 1958, and note.

According to Squire (1987), the hippocampus also receives a diffuse projection from the lateral hypothalamus.)

In the hippocampus itself, high circulating glucocorticoid levels reportedly cause down-regulation of Type 1 and 2 receptors, although, not unexpectedly, "the relative magnitude of the change" was greater in Type 1 (McEwen et al., loc. cit.). This is apparently a homeostatic response but not always an effective one. For since down-regulation has a nonzero lower limit, prolonged glucocorticoid exposure can result in the death of hippocampal neurons, seemingly by a neurotoxic saturation effect (cf. Sapolsky and Pulsinelli, 1985).

McEwen et al. (loc. cit.) conclude that "the hippocampus shows an apparent sensitivity to neuropeptides such as ACTH. This sensitivity is not evident in the cerebral cortex.... There are indications that circulating ACTH and vasopressin influence functions of the blood brain barrier, and that these are more pronounced in the hippocampal region.... *Current information concerning glucocorticoid effects in the brain indicates that the hippocampus is a unique hormone target area....*" (Italics added.)

This sensitivity to ACTH and vasopressin, coupled with the presence of both high-affinity Type 1 and low-affinity Type 2 glucocorticoid receptors (and the ability of Type 1 to mediate rises in synapsin levels), are features that gear hippocampal output to the concurrent level of arousal, over a range running from mild alerting to massive activation. It is this *range* of responsiveness that makes the hippocampus figure in a range of memory functions, from transient "primacy" effects to those underlying indefinitely long recall.

Nor does this monitoring function affect only its output to the cortex; the hippocampus has also been reported to act, via the hypothalamus, to trigger the brainstem release of β-endorphins, as a sequel to "novelty" responses, inescapable foot shock, or during passive or active avoidance training (Izquierdo and Netto, 1985; rat).

Taken together, these findings suggest why it is that the hippocampus has been found to have a central role in long-term memory formation. The time relations here are worth noting. For since glucocorticoid output, especially following acute arousal, is likely to remain above normal (waking, resting state) levels well past the peak of the initial stress-producing episode(s), much of the hippocampal activity leading to LTM formation may also occur in this poststress interval.

Thus LTP, established in the FD and hippocampus during the rising or peak phases of central activation, will tend to continue on into the "downslope" phase (Fair, 1988, p. 168); and it is in this phase that circular interaction between the hippocampus and neocortex is likely to

A study by Weingartner et al., 1981, reported that eight normal subjects, treated for two weeks with an intranasally administered vasopression analogue, were able to "learn information more effectively."

De Wied and De Kloet (1987) state that "the effects of ACTH-related peptides are of short term (attention, motivation, learning) and vasopressin with long-term...control of adaptive behavior" (via long-term memory).

Brattleboro rats, which have a genetic deficiency in AVP (vasopressin) synthesis, also "suffer from severe memory deficits" (De Kloet et al., 1987).

Neurons of the parvocellular paraventricular nucleus, containing both corticotropin releasing factor (CRF) and vasopressin, project to the zona externa of the median eminence, where they act via the portal system to regulate ACTH output of the anterior pituitary. Zimmerman (1987) reports that vasopressin potentiates the effects of CRF.

These data suggest that vasopressin may have the longer-term effects described above, essentially by this mechanism—that is, by prolonging, or adding a tonic component to, the initial phasic activation of the hypothalamic-pituitary-adrenal (HPA) system. As described, *passim*, in the text here, such prolongation may be one of the conditions conducive, or even necessary, to LTM formation.

result in LTM formation. (See below, Chapter 5, Part 3.)

As will be seen later in the text, these relations underscore the fact some minimal rise in nonspecific or reticulocortical input may be essential to the whole process. For it is this nonspecific component that may provide the "coactivation" leading to long-term potentiation, whether or not the units involved also show a Hebbian convergence of specific inputs, and whether or not they are in the hippocampus or neocortex (Chapter 4, Part 2).

Entorhinalis and the perforant pathway are key structures in the bidirectional interface between the hippocampal system and neocortex. Fallon and Loughlin (1987), in a detailed analysis of the forebrain projections of what they call the SN/VTA system (the ventral tegmentum and substantia nigra, pars compacta), report that whereas piriform cortex receives projections from both the SN and VTA, entorhinalis receives projections only from the VTA and from "scattered cells in the retrorubral field."

They also point out that "whereas other VTA and SN neurons collateralize to prefrontal and suprarhinal/perirhinal/piriform cortices . . . projections to entorhinal cortex are from neurons that do not."

These features may be extremely important; specifically, they suggest that "motivational" input to entorhinalis is from the basal dopaminergic system, without an extrapyramidal component (from SN), and without diffuse distribution of the same input to other areas (via collaterals). This may, in other words, be an arrangement that acts, at the level of subcortico-cortical input, to separate inputs conducive to the "declarative" (or sensory-related) type of memory formation from those conducive to the more frontally-represented "procedural" (or action-related) types of memory.

In the preceding sections, I have described some of the cytoarchitectonic features that, in the relay of sense-data from prime receiving to association areas, favor the several stage invariance sorting that leads to generic recognition. The thinning of layer IV that occurs as one moves from sensory and parasensory to association cortex implies a shift from relay of highly fractionated data (such as are received by IV in area 17), to relay of data that have already been processed and somewhat condensed or generalized.

In other words, as the parvocellular (layer IV) component diminishes and the magnocellular (lower III) component increases in columns across the cortex, the effect is to increase the generalizing function of the relay and decrease the resolution or particularity of the data relayed. This process is directly reflected in an electrophysiological study by Quirk and Muller (1989), who report that in the rat, visual unitary responses in entorhinalis were "more tightly controlled by the cue-card"—i.e., more specific—than were the responses of hippocampal units. The latter occurred later and were more holistic, being "dramatically" affected by changes in the shape of the viewing chamber.

It should be recalled that some of the detail lost in this generalizing process survives at earlier stages in the relay, and so is retrievable by playback. (The latter is not to be taken literally as meaning point-for-point reverse transmission in the sensory throughput network. It probably involves the return pathways from entorhinalis to the several cortical areas described in Chapter 4—a form of feedback proposed here to figure in the "consolidation" of long-term memories or their subsequent retrieval.)

Consider, now, the assemblies through which incoming sense-data are passed in final stages of the relay from prime receiving areas to the dentate gyrus and hippocampus. In entorhinalis (areas 28a and 28b) the cortex becomes four-layered, with the partially empty *lamina dissecans* between the bottom two. In the presubiculum, it becomes two-layered, and in the subiculum and hippocampus, one-layered.

Functionally, the arrays of large pyramids in CA3–CA1 may represent the "integrating" action of magnocellular bands, such as those in cortical lower III, taken to an extreme. It is then the end results of this form of data compression that are relayed back, via the perforant path, to far distal cortex where, by the unitary mechanisms suggested in Chapter 4, they act to establish what could be called core elements in LTM.

These might be defined as constellations of highly generalized key features, extracted from immediately precurrent sensory inputs by transmission over the pathways just described. They are core elements in the sense of providing the base, in neocortex, from which, by playback, the remaining components of a given long-term memory then become incorporated into it.

The latter process, equivalent to what has been called consolidation, occurs over a period of minutes to days, and extends from far-distal association cortex to areas or assemblies approaching, but probably not including, areas such as 17–18. (For some of the interareal routes involved, see Fig. 4-4 below.)

Squire notes that this primary form of consolidation may lead on to secondary forms, by which initially separate items in LTM become associatively interconnected. The "core" elements in such memories may owe their persistence to the fact that, unlike those described by Blakemore, they result from postsynaptic membrane changes that, for reasons given in the next chapter, tend to be indefinitely long lasting. They thus do not depend upon day-to-day inputs for maintainenance of a particular response bias. Because their core elements represent particular selections from experience, which are then highly resistant to change over time, memories of this type, once consolidated, form the basis of specific (as opposed to generic) recognition. In effect, this account leads to a prediction that appears to be borne out by data reported in the neurological literature.

Damasio (1985) states that in prosopagnosia, "the notion that the deficit ... is limited to human faces is erroneous; the deficit is of far greater magnitude.... Patients with prosopagnosia can recognize any object in the environment, provided the examiner does not require a recognition of the specific object within the group.... In other words, these patients can perform a *generic recognition* ... but the identification of a *specific member* within the generic class eludes [them]" (italics original). The patient visually recognizes people as people, cars as cars, or cows as cows, but cannot visually identify them if they happen to be *his* (relatives, cows or cars).

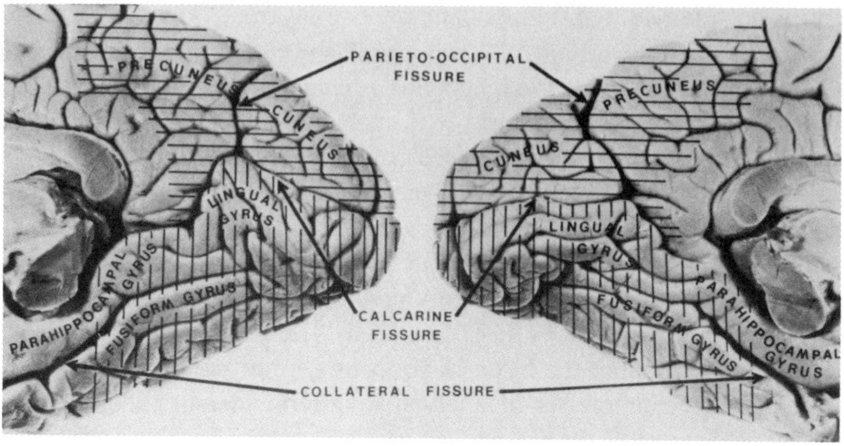

Figure 3-1. Mesial view of human posterior cortex. Vertical hatching indicates areas in which bilateral lesions can produce visual agnosias including prosopagnosia. Horizontal hatching indicates areas in which bilateral lesions can result in Balint's syndrome. Reprinted with permission of FA Davis Company, from Mesulam, M-M (1985): *Principles of Behavioral Neurology*. Philadelphia: FA Davis Company.

(See Tranel and Damasio, 1985, and Appendix II, below; also Pillon et al., 1987, who report a case in which visual recognition of familiar faces was selectively impaired following right hemisphere lesions only—in support of a number of studies indicating right hemisphere dominance for that particular function.)

According to Damasio, "all cases of prosopagnosia in which the defect endured beyond the acute period proved to have bilateral lesions. The lesions compromise either the inferior and mesial visual association cortices (in the lingual and fusiform gyri) or their subjacent white matter." (See Figure 3-1.)

In these patients, the routes of relay have been cut bilaterally so as to prevent incoming visual data from accessing the core elements in LTM to which they may correspond. The cut involves a zone of visual association cortex evidently transitional between specific and generic visual memories, being so placed as to leave the latter intact. Homologously placed cuts, in animals shown to be capable of highly specific visual recognition, should experimentally duplicate the results Damasio reports. (In such experiments, olfactory as well as auditory inputs would have to be excluded.)

In man, when damage to visual association cortex is "comparable but more extensive," e.g., when it involves "bilateral lesions in the ventral and mesial parts of the occipitotemporal visual areas that often extend dorsally and laterally," the result is the condition known as visual object agnosia. Damasio reports that "in addition to prosopagnosia, these patients have an inability to recognize even the generic class to which an object belongs"—a finding that implies that visual memories responsible for generic recognition are established in cortex lying closer to the primary visual area.

Interestingly, Damasio reports that some of the patients with visual object agnosia may "unblock," becoming able to (generically) identify an object when it is moved or slowly rotated. In such patients, a part of the visual system, described by Livingstone and Hubel (1988) as related to motion detection may still be intact, permitting generic recognition by that route. (The path would run: magnocellular layers 5–6 in LG → 4B in 17 → "thick stripes" in 18 → "MT" (middle temporal lobe) → parietal cortex; Hubel, 1989.)

Corbetta et al. (1990; see also Appendix I, below) report a study, in which positron emission tomography was used to record focal cerebral bloodflow (CBF) increases in 9 human subjects during tests requiring them to attend to changes in the shape, color or velocity of a complex stimulus. Attention to velocity elicited CBF increases in an area of the left inferior parietal lobule identified by the authors as Brodmann 39. Shape responses also occurred unilaterally (on the left) at a site in the mesial occipitoparietal sulcus. Those, and other results of this study, are summarized in Table 3-2. Note the distribution of shape responses.

The point is that the agnosias Damasio describes and the loci of the lesions that appear to produce them are consistent with the model of cortical memory functions proposed here.

According to that model, the flow of information across the cortex, from the prime receiving into the association areas, and finally to entorhinalis, entails increasing generalization of the data relayed. Since, at the several stages in this throughput system, the data also become associatively cross-linked (Fig. 2-2, above), they are more holistic. That is, because each of the cross-linked components is itself generalized, the linkage between them amounts to a higher order generalization, with a proportionately broader base in specific referents represented at earlier stages in the relay.

("Association" among elements at a given level in the relay amounts to an increase in the probability of their concurrent or immediately sequential activation. Association, involving assemblies, is thus a second order consequence of unitary response preferences established on Blakemore's principle.)

Table 3-2.

Shape	Color	Velocity
Lingual gyrus (L)	Lingual gyrus (L+R)	—
Fusiform + para-hippocampal gyri (L+R)	—	—
"Midway along superior temporal sulcus."(L+R)	—	—
—	Strip in dorso-lateral 19 (L+R)	—
Mesial parieto-occipital sulcus (L)	—	—
—	—	Area 39 (L)

Cortical areas in which sites showed bloodflow increases during attention directed to changes in the shape, color, or velocity of a compound stimulus. 9 Ss; data from PET scan. (Corbetta et al., 1990. See Appendix, Note 1.)

Activation of sites in the fusiform and parahippocampal gyri during attention to shape suggests what, in the text, is termed "registration"—involvement of the hippocampus and return projections from entorhinalis. When followed by consolidation, it results in LTM formation; when not, it leads to short-term retention as in the primacy effect.

As explained in the text, the primacy effect is essentially an onset phenomenon, a form of "novelty" or "orienting" response. Maintenance of attention, once a new stimulus train has been initiated, is more probably a frontal function. Milner's hippocampectomized patient Henry, if not distracted, could sustain attention to an assigned task. He did not show the primacy effect.

For cytoarchitectural reasons, this form of data condensation probably reaches a maximum over the trisynaptic path: entorhinalis → FD → CA3 → CA1 → entorhinalis. The path includes a shortcut, from 28b direct to CA1, which may be of special importance since 28b receives input from prepiriform cortex (area 51) and orbitofrontal areas 12 and 13 (macaque; Van Hoesen et al., 1972).

This path may consequently be involved in fast-acting global responses such as those mediated by the amygdala; and whether facilitatory or inhibitory, its output could have important override functions with respect to CA1 inputs arriving via the longer route from FD.

The return pathways from entorhinalis thus convey highly condensed, generalized data some of which may then become "fixed" (by the dendritic changes discussed in Section 4) in assemblies of mesial or basal association cortex, forming what I have called core components of LTM. With respect to prime receiving areas, these assemblies are *distal* to those mediating generic recognition, meaning that in vision, for example, we first perceive what something generically is before specifically identifying it (cf. Ambros-Ingerson et al., 1990). If this is the actual arrangement, it would lead to the clinical findings reported by Damasio.

From a psychological standpoint, this model is paradoxical, in that it says our most enduring specific recollections originate not in the most specific but in the most highly generalized contents of memory. And whereas we tend to think we generalize from specific instances accumulated in long-term memory (and can, on occasion, do so), the basic process is anterior to that and effectively preconscious.

Simply by the way the cortical relay systems are organized, they generalize from the specific thalamocortical sensory inputs that are part of our daily experience, thereby providing us with an awareness of our surroundings whose categories seem readymade. This model would, in fact, predict that focal lesions to posterior cortex or the underlying white matter might result in category-specific deficits of the type described by Nielsen (1946) or McCarthy and Warrington (1988; for a review of this literature see Shallice, 1988.) The findings of Corbetta et al. (1990), shown in Table 3-1, suggest why color coding has proved so useful as a mnemonic aid, or conversely, why posterior cortical lesions leading to difficulties in visual recognition are sometimes accompanied by achromatopsia. (Cf. also, Livingstone and Hubel, 1988, on the distribution of color-sensitive units in visual prime receiving cortex.)

To summarize: cortex of the lingual and fusiform gyri in man evidently forms a watershed area in the sense that local lesions or interruption of the underlying fiber tracts can cut the routes leading from generic to specific visual recognition (as in prosopagnosia). Note that, as shown in Figure 3-1, this is mesial cortex, transitional to entorhinalis.

Over this route, during LTM formation, incoming sense-data are referred to the FD-hippocampal system in which some of these data may then be selected for incorporation into long-term recall. (Figure 3-2, compiled from data reported by Jones, 1974, indicates the order in which activity in the limbic system is likely to be entrained during input to the visual cortex.) Selection will depend upon whether the level of activity in the "motivational" systems at the time is sufficient to produce conditions making for long-term potentiation (LTP; see next section). Here, dopaminergic input to the hippocampus over pathways from the VTA described above may play an important part. This is the stage of memory formation referred to here as "registration."

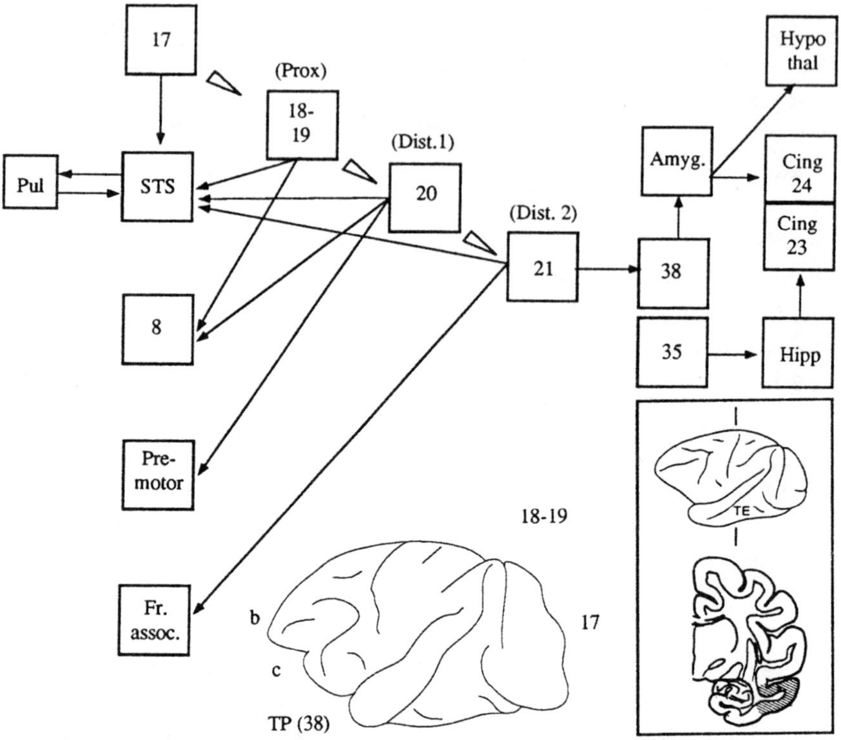

Figure 3-2. The pathways shown in the schematic, and some of the corresponding areas in macaque cortex (bottom) are taken from Jones (1974). Abbreviations: Amyg, amygdala; Dist. 1, Dist. 2, distal association areas 1 and 2 (see below); Fr. assoc., frontal association cortex; Hipp, hippocampus; Hypothal, hypothalamus; Prox, proximal association cortex (see below); Pul, pulvinar; STS, posterior bank, superior temporal sulcus; TP, temporal pole.

Proximal and distal refer to Graybiel's division of association areas, according to their distance from prime receiving or frontal motor sectors, and their subcortical outflows to the limbic system (Graybiel, 1974).

The inset is from Fuster and Jervey (1981). The upper panel shows the level at which the frontal section below it was made. The shaded section in the lower panel shows the inferotemporal cortex from which these authors obtained single unit recordings from color-sensitive cells. Recordings were made from 500 cells in nine macaques, during a delayed matching-to-sample task in which a color or color and symbol combination was the cue for a food reward.

The authors concluded that "a substantial number of cells showed color-dependent differences in frequency of discharge during the retention periods of the task." The results appear to support the speculation by Gross et al. (1969; 1974) that "adequacy of visual stimuli for driving IT cells might depend on the significance of the stimuli as well as on their physical characteristics."

Inset reprinted with permission of the AAAS, from Fuster, JM, and Jervey, JP (1981): "Inferotemporal Neurons Distinguish and Retain Behaviorally Relevant Features of Visual Stimuli." *Science* 212: 952–955.

During it, playback via the perforant path acts to establish potential "core" elements of an LTM, e.g., in mesial cortex such as that of the fusiform gyrus. This represents a preliminary or preparative stage of consolidation. (See Table 3-2 and caption, above.) For the final stage, or consolidation, to occur, will then require repetitive playback over efferents (such as those from entorhinalis, mentioned above) by way of which the generic components of a given memory become recursively linked to its "core" components. Because dependent upon circular iteration, the process is time-dependent as described below.

The result is a set of pathways that can be accessed from either direction—from prime receiving areas in, mediating generic or specific recognition of immediate sense-data (in that order); or from mesial cortex out, mediating "retrieval," or specific recall not necessarily corresponding to any currently incoming sense-data.

If this system of connections is cut, e.g., at the level of the lingual and fusiform gyri in man, specific (visual) recognition will be impaired but generic recognition will survive. If the lesions are expanded to include the additional visual cortex described by Damasio, both forms of recognition will be impaired.

An early study by Flexner et al. (1963) appeared to show that consolidation in the mouse occurs "backward"—that is, by "irradiation of the trace" from an area homologous to temporal cortex, back towards the prime receiving areas, a process lasting from 5 to 11 days. Impairment of protein synthesis by puromycin reportedly blocked consolidation, possibly by blocking the action of the immediate early gene zif/268 and its mRNA, which have recently been found to figure in long-term potentiation (Cole et al., 1989; see inset note on next page). Squire (1987, pp. 146–148) reviews evidence that inhibition of cerebral protein synthesis, e.g., by cycloheximide, blocks LTM formation (consolidation) without significant effect on short-term learning, e.g., during training.

Finally, the two stages of long-term memory formation just described are experimentally separable. As shown by Milner's patient Henry (1963) or Squire's patient R.B. (1986), bilateral damage to the hippocampus can block the whole process, making further additions to LTM difficult or impossible. (Cf. the long-term follow-up study on Henry, reported by Corkin et al., 1989.)

If the hippocampal system is intact, consolidation can still be impaired or prevented, for instance by electrocortical shock or anesthetics delivered at a suitable interval after exposure of an animal to a conditioning stimulus. Thus, Abt et al. (1963) showed that, in mice, retention of a conditioned response established by one-trial exposure to foot shock

This gene is reportedly activated in the suprachiasmatic nucleus of rats and hamsters, its mRNA showing "dramatic increases" when, during their normal dark period, the animals received 30 minutes' exposure to light (Rusak et al., 1990).

What makes this finding of particular interest is that it involves the phylogenetically old retino-hypothalamic pathway and follows an "orienting" or "novelty" response pattern. That is, it is triggered by a break in an habitually expected input condition (continued darkness).

Events leading to LTP and registration via the FD and hippocampus follow essentially the same pattern, the simplest example being the primacy effect. More complex examples would include external events having "surprise" value for the basal motivational system—e.g., inputs causing abrupt changes in drive or affect. In all cases, the critical variable is the concurrent rise in central activation, since that may be what provides the "coactivation" needed to mobilize this gene and its mRNA.

This interpretation would predict that in the Rusak study, pretreatment of the animals so as to diminish the arousal effect of unexpected light exposure (as monitored for instance in the EEG) would also diminish or abolish activation of this IEG (termed zif/268 by Cole et al., or NGFI-A by Rusak et al.).

was abolished by ether administration 8 minutes afterward, impaired by administration at 16–20 minutes, and unaffected if ether was given at 24 minutes or later.

In the chapter that follows, I will discuss unitary mechanisms that recent studies have suggested may figure in the types of memory formation outlined above. I will also propose the differential roles that two of these mechanisms may play in the processes equivalent to consolidation.

Chapter 4
Fundamental Mechanisms

Part 1. Long-term Potentiation

Abstract. Discussion in detail of the mechanisms involved in long-term memory (LTM) formation at the level of single neurons. Similarities between sensitization and long-term potentiation (LTP). Evidence that LTP is presynaptic. Critique of Alkon's "dendritic compartment" model of postsynaptic learning. Possibility of positive feedback at the synapse during learning. Intracellular dendritic transfer of protein kinase C (PKC); its possible relation to glutamate receptor types, LTP, and Hebbian convergence effects.

In recent years, an immense amount of work has been done on the processes thought to underlie memory; but because of differences in approach and terminology, some essential similarities between the various findings may have been overlooked. For example, sensitization as defined by Kandel and Schwartz (1985; in *Aplysia*) has several features in common with long-term potentiation (LTP), as described, for instance, by Brown et al. (1988).

Both are "associative" and appear to involve a voltage- or a receptor-dependent gating of Ca^{2+}, in which a Ca^{2+}-dependent protein kinase may play a critical role. (An exception may be the LTP elicited in CA3 hippocampal units by mossy fiber stimulation, which is reportedly not voltage-dependent, and not blocked by NMDA receptor antagonists or by intracellular calcium chelators, but does appear to depend upon spiking frequency (Zalutsky and Nicoll, 1990). In the same system, Hopkins and Johnston, 1984, showed that bath-applied norepinephrine potentiated this effect.)

Sensitization may take short- or long-term forms; depending upon the conditions inducing it, LTP also shows variable duration on a scale of hours to weeks. Sensitization, e.g., in *Aplysia,* is reported to depend chiefly upon presynaptic changes that increase transmitter output of the sensitized unit. The long-term form reportedly involves increases in the number of vesicles and in the area of "active zones" at axon terminals (Bailey and Chen, 1983)—changes that "might require the expression of new genes" (Kandel and Schwartz, loc. cit.).

It was subsequently shown that mRNA inhibitors selectively block long-term sensitization (Montarolo et al., 1986). In LTP, it has been found that concentrations of the mRNA of zif/268—one of four immediate early genes (IEGs) whose mRNAs can be induced in the brain by seizures—are sharply increased in dentate gyrus granules by high-frequency (but not by low-frequency) stimulation of the perforant pathway (Cole et al., 1990; in vivo, rat).

A distinguishing feature of sensitization is that, being primarily presynaptic, it tends to generalize—that is, to produce corresponding changes in the output of the units to which it projects (whose numbers, in a system such as neocortex, can be very large). Kandel and Schwartz (loc. cit.) describe sensitization as resulting, behaviorally, in an animal that tends "to respond vigorously to a variety of previously neutral or indifferent stimuli after it has been exposed to a potentially threatening or noxious stimulus.... Whereas habituation is limited to the stimulated pathways, sensitization has a more widespread distribution, involving both stimulated and unstimulated pathways."

In single units, sensitization resulting, e.g., from Hebbian convergence, or "coactivation" by inputs from more than one source, may not result in increased responsiveness specifically to those inputs. Rather, the probability that the unit will fire in response to any of its inputs is increased.

The critical change, in other words, has occurred, not at particular sites in the cell's receptor membrane, but in the mechanisms that determine its threshold generally, and (in the case of long-term sensitization) its vesicular output at axon terminals. The cell, in short, does not "remember" the configuration of inputs that caused it to become sensitized, as would be the case if the process included site-specific changes in its receptor membrane making it differentially responsive to those inputs.

I have spelled this point out because there has been some disagreement as to whether LTP is presynaptic or postsynaptic, and some confusion as to what role it may play in memory formation. Here I have taken sensitization as an instance of presynaptic learning. The recent quantal studies by Malinow and Tsien (1990), and Bekkers and Stevens (1990) seem to show that long-term potentiation (LTP) is also presynaptic, which raises the question of how it can lead to the specificity typical of long-term declarative memory.

Postsynaptic models of memory formation such as Lynch and Baudry's (1985) suggest a solution to this problem. The essential feature of such models is that they involve site-specific changes in dendritic membrane of the postsynaptic cell. Such changes make a given unit preferentially responsive to certain of its inputs; and in the aggregate—in a given assembly or in those to that it projects—they amount to the formation of a subset of preferred pathways, selected from a set of *N* possible pathways.

In their 1985 model, Lynch and Baudry proposed that bursts of high-frequency input caused an influx of calcium into dendritic spines, activating a calcium-dependent proteinase that then uncovered "occluded" glutamate receptors. Prolongation or repetition of these input-conditions might then so alter local membrane structure as to produce "morphological changes [that] could last for very long periods" (Lynch and Baudry, 1985).

Such changes, amounting to increases in synaptic efficiency at certain postsynaptic sites, and therefore corresponding to certain configurations of presynaptic input, might provide a basis for long-term memory formation. According to evidence reviewed below, LTP, although presynaptic, might involve such site-specific postsynaptic changes. However, like LTP itself, these may not be of sufficient duration to support LTM. The Bekkers and Stevens, and Malinow and Tsien studies just cited suggest that LTP increases the probability of a unit's "firing" (or decreases its probable "failure rate").

This raises two questions: (1) Are these changes in any way selective for the upstream inputs that induce them? (2) What effects do they have downstream?

Alkon (1989) has outlined a two-stage model of LTP, in which the second stage, involving a delayed transfer of PKC to dendritic membrane, might add the required element of specificity. Two other studies provide data relevant to his model. I will review these first, and Alkon's afterwards.

Malenka et al. (1989) present evidence, obtained in vitro from hippocampal slices, that a calcium/calmodulin-dependent protein kinase (CaM-KII) figures in the first stage of LTP. Their data "indicate that activation of postsynaptic CaM, presumably by the rise in dendritic spine Ca^{2+} initiated by tetanic stimulation, is a requisite step in the induction of LTP. Furthermore we have demonstrated that postsynaptic kinase activity is also required for LTP."

However, they note that the synthetic CaM-binding peptide CBP, which inhibits LTP when applied intracellularly in vitro, does not inhibit PKC activity in the same preparation, suggesting that "PKC alone cannot be responsible for LTP."

Malinow et al. (1989) conclude that both PKC and CaM-KII may be necessary to LTP, and that "both are activated after Ca^{2+} entry through

NMDA channels." They add that "these kinases could act in parallel, although the finding that postsynaptic injection of PKC results in enhanced transmission suggests that they act in series."

This conclusion receives inferential support from the results reported by Alkon (1989). He found that in CA1 cells in the rabbit, during conditioning, concentrations of membrane-associated PKC initially increased in the somata. However, "three days after conditioning, distribution of the PKC label was entirely transformed: labeling is enhanced in the dendritic regions to a much greater extent than it is in the cell bodies"—a change he suggests may result in "an association [being] stored in particular dendritic compartments." (For a similar, but more generalized model of postsynaptic learning, see Alkon, 1984.)

These time relations are obviously important, if we assume that LTP occurred in CA1 units during conditioning of the rabbits in this study. They imply that PKC, besides its involvement in the first (onset) stage of LTP, may play a critical role in its subsequent maintenance phase, and also (as Alkon suggests) in its specificity.

He proposes that the transfer of PKC to dendritic membrane could, in effect, conserve a given memory "without saturating the capacity of CA1 cells to store additional information." As was proposed for hypothetical neurons by Poggio and Torre (1981), this would give an assembly of such units a storage capacity proportioned to the number of their synaptic inputs, making it several orders of magnitude larger than it would be if proportioned to the number of units in the assembly.

This is an attractive model that, however, involves certain difficulties. The data do not tell us whether PKC was specifically concentrated at the sites in dendritic membrane by way of which LTP had been induced. If it was, Alkon's conclusion might follow. If not, if the transfer of PKC was a general one, involving the dendritic apparatus nondifferentially, his conclusion might not follow, since no "dendritic compartment" would be favored over any other as a potential route for activation of the unit. The neuron's probability of response to any adequate set of inputs would simply have increased, as appears to occur in sensitization.

There is a second difficulty, for even if we assume that the transfer of PKC occurs preferentially to certain dendritic sites—those representing the LTP-generating input—what would be the half-life of the phenomenon? How long would those "dendritic compartments" continue to make the neuron preferentially responsive to that input pattern, and thus (as Alkon seems to imply) part of a preferred pathway or memory?

Some answers to these questions are suggested by a strikingly similar biochemical mechanism, described by Rasmussen (1989). He outlines the sequence of intracellular events touched off by angiotensin II via its receptor in an adrenal glomerulosa cell. Like LTP, the process is divisible into overlapping but "temporally distinct phases," with calcium-calmodulin playing a key role in the first phase, and membrane-localized PKC in the second. (See Figs. 4-1 and 4-2.)

The resulting "memory" effect is illustrated in Fig. 4-2. "If isolated adrenal cells are perfused... with angiotension II for three periods of between 15 and 20 minutes, separated by intervals of similar duration, the aldosterone secretion that occurs during each successive exposure is higher than... during the preceding one.... That memory is transient." That is, as intervals between perfusions are made longer the effect is attenuated or disappears.

Benowitz and Perrone-Bizzozero (1989) report that GAP-43, one of a group of neuronal polypetides related to axonal growth or increases in synaptic efficacy, is found in certain subpopulations of the nervous system in rats and humans. They state that "GAP-43 is a substrate of (PKC); when extracted from the membrane it binds calmodulin in the absence of calcium in the membrane." It "appears to be important in regulating" the phosphorylation of phosphatidyl inositol to PIP2, setting up the sequence whereby phospholipase C generates DAG and IP3. In other words, in the sequence shown in Fig. 4-1, it would occur secondarily to the initial PIP2 $\rightarrow$ PIP3 + DAG reaction and therefore act as a positive feedback.

These authors in fact state that "PKC-mediated changes in the phosphorylation of GAP-43" [have been found to] "correlate highly with the persistence of long-term potentiation (LTP) in the rat hippocampus."

GAP-43 is a protein that figures in regeneration (e.g., of the cut optic nerve in goldfish) or in early postnatal development of the optic and pyramidal tracts. Maximal concentrations of its mRNA coincide "with the period of maximal axonal outgrowth and synaptogenesis, and then fall off abruptly."

Interestingly, in the adult brain, it continues to be expressed in areas corresponding to phylogenetically old parts of the system. These areas include layer I of rat neocortex (that layer being the rostralmost terminus of the reticular system), the inner third of the molecular layer in FD, hippocampal CA1 cells, pyriform cortex and "in a continuum of subcortical structures including caudate-putamen, portions of the amygdala," and parts of the thalamus not included in the specific projection system.

In the human brain, the highest levels of GAP-43 mRNA are found in layer II in cortical association areas (some of which, by phylogeny, are quite old) and in hippocampal pyramids. Levels "are very low, by contrast, in primary sensory and motor areas of the cortex, in cerebellum, and

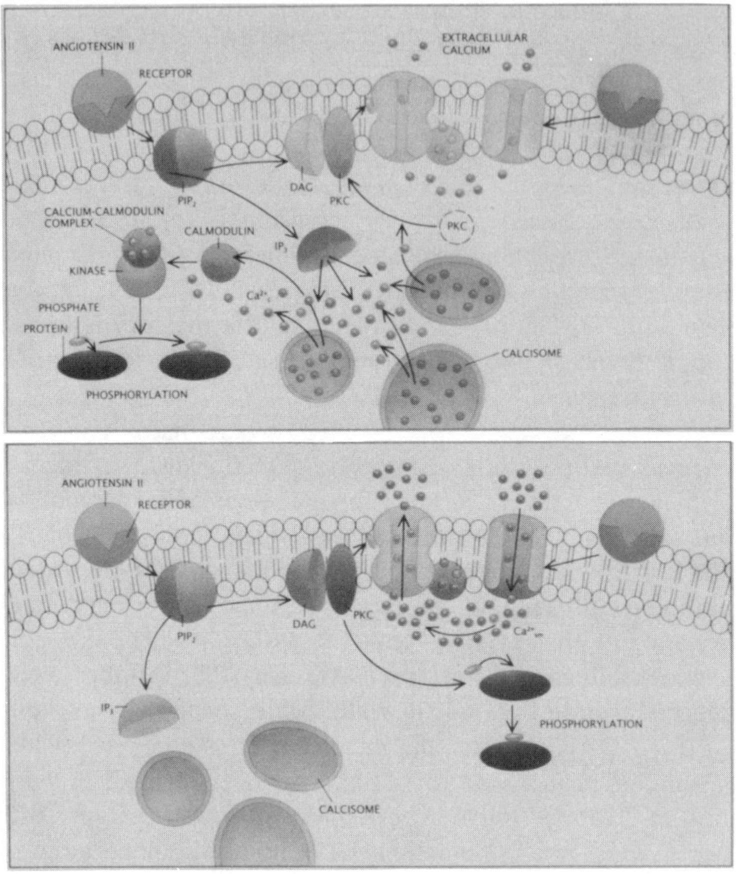

Figure 4-1. The upper panel of the schematic shows the sequence of intracellular events when angiotensin II activates its receptor in an adrenal glomerulosa cell. PIP2 (phosphatidyl inositol diphosphate) is broken down into IP3 (inositol triphosphate) and DAG (diacylglycerol). IP3 causes the release of calcium ions from calcisomes into the cytosol. Some of the ions bind to calmodulin, producing a complex that activates protein kinases, which in turn phosphorylate the proteins mediating the cell's response (secretion of aldosterone). Release of Ca^{2+} from the calcisomes and the increase in DAG also stimulate the transfer of PKC (protein kinase C) from the cytosol to the cell membrane, a shift similar to the one reported by Alkon to occur in rabbit CA1 cells during conditioning.

During the sustained phase of the reaction, (lower panel), angiotensin II increases calcium cycling across the membrane. The local (submembrane) increase in Ca^{2+} concentration activates the membrane-associated PKC, which phosphorylates a second set of proteins that sustain aldosterone secretion. This again suggests an analogous mechanism that in neurons might act recursively to enhance activity at the synapse during LTP.

Reprinted with permission of Scientific American, from Rasmussen, H (1989): The Cycling of Calcium as an Intracellular Messenger. *Scientific American*, 261: 66–73.

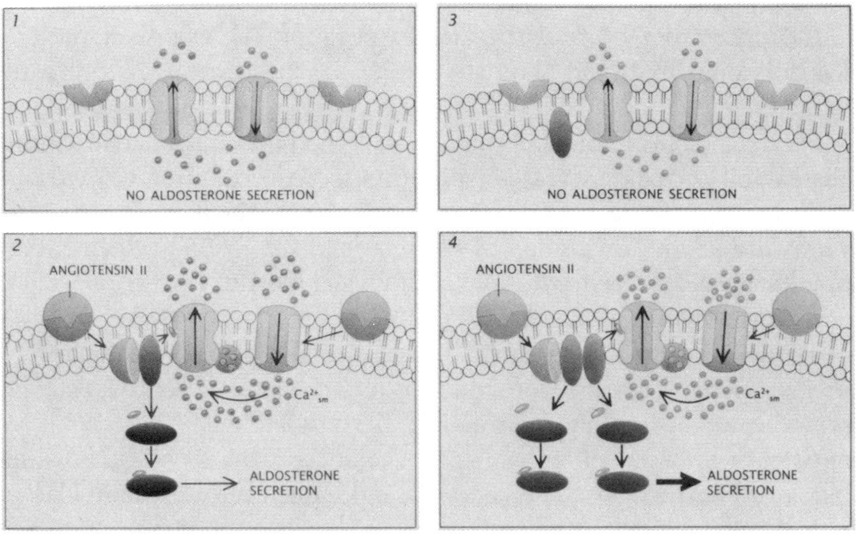

Figure 4-2. Shows the holdover or "memory" effect when adrenal glomerulosa cells are exposed to successive doses of angiotensin II in vitro. Panel 1 represents the resting state. The rate of calcium cycling is low and PKC is not localized in the membrane. Panel 2 shows some of the cascade depicted in Fig. 4-1. In panel 3, angiotensin II is absent but PKC remains in the membrane. When the cell is re-exposed to angiotensin II some minutes later (panel 4), PKC transferred from the cytosol to the membrane is added to that already there, and aldosterone secretion and calcium cycling are correspondingly increased.

Reprinted with permission of Scientific American, from Rasmussen, H (1989): The Cycling of Calcium as an Intracellular Messenger. *Scientific American*, 261: 66–73.

in most other parts of the brainstem" (Benowitz and Perrone-Bizzozero, loc. cit.).

These findings suggest that, unlike the nerve growth factors discussed in part 2 of this section, GAP-43 may have chiefly presynaptic effects, involving axonal changes that may increase the output of a given neuron across the board, or without differential effects on units downstream.

The question is then whether some of the foregoing mechanisms might not account for the dendritic transfer of PKC reported by Alkon to have been found in rabbit CA1 neurons three days after conditioning. The second question is whether this transfer might not, as in adrenal glomerular cells, serve as a pump, recycling CA^{2+} during activation of the neuron, thereby setting up a positive feedback at the synapse. In other words, if PKC transfer were site-specific, could it also have differential effects at synapses upstream, resulting in a pro tem link, equivalent to a memory for the corresponding input(s)?

Evidence from other tissue studies suggests (1) that such positive feedback loops exist and (2) that some of the mechanisms involved are intracellular, and more complicated than the ones just outlined.

For example, Harootunian et al. (1991) report that "many nonexcitable cells exhibit periodic increases (spikes) in the concentration of cytosolic free calcium... when stimulated with hormones [such as vasopressin] or growth factors." In a study of fibroblasts of the REF52 cell line, they found that oscillations in cytosolic calcium and inositol 1,4,5-triphosphate (IP3) did not require "functional" PKC.

Their data led them to conclude that elevated cytosolic calcium "by itself is neither necessary nor sufficient to increase IP3, yet it synergizes with receptor occupancy to stimulate IP3 production." The results are consistent with a model "in which the major feedback loop is Ca^{2+} stimulation of phospholipase C to generate IP3, which releases more Ca^{2+} if the internal stores are sufficiently full.... The plethora of G proteins and isozymes of phospholipase C suggest that the interaction with [cytosolic calcium] is likely to vary from tissue to tissue."

(They add that it is "currently controversial whether phospholipase C is significantly stimulated" by increases in cytosolic Ca^{2+} and "if so whether the stimulation is in parallel to or synergistic with receptor-G protein activation.")

The critical question, concerning Alkon's interpretation of the membrane transfer of PKC that occurred during the conditioning of his rabbits, comes down to this: was the transfer site-specific or general? If it were general, affecting all input sites, most frequently "seen" dendritic inputs would not become preferred. There would be no differential effect—effectively, no memory formation.

Data reviewed in chapter 5, part 2, below indicate that there are a number of glutamate receptors, including four fast-acting subtypes, each of which has two forms derived from alternative gene splicing. If we assume that two structurally different receptors for the same transmitter correspond to different sites on receptor membrane of a given neuron, inputs via those receptors would by definition be site-specific. One would also expect them to be additive, as appears to be the case.

Davies et al. (1989) note that whereas induction of LTP in hippocampal CA1 units by tetanic stimulation involves N-methyl-d-aspartate (NMDA) receptors, the quisqualate receptor "mediates the fast synaptic response evoked by low-frequency stimulation."

They report that, in hippocampal slices following induction of LTP (in CA1 units, by stimulation of the Schaffer collaterals), iontophoretic application of the quisqualate receptor ligand AMPA (a-amino-5-Me-4-

isoxazole propionic acid) produces a delayed increase in quisqualate-induced depolarization—the question being what caused the delay.

The increase, on the order of 20%, began ~15 minutes post-LTP, reaching a maximum in four cells at 25–30 minutes, and in a fifth cell at 1 hour. If quisqualate receptor activation was repeated 4 times at 15-minute intervals preceding LTP-induction, the immediate post-LTP quisqualate response was unaltered but potentiation of the delayed response was extended, reaching a maximum at 2 hours and lasting for 3.

Davies et al. add that in the presence of PKC inhibitors, "LTP lasts for only 1–2 hours.... Moreover, injection of purified PKC into the postsynaptic neuron induces an LTP-like change in synaptic transmission."

Taken together, these findings imply that PKC transfer (from cytosol to somatic and finally to dendritic membrane) is maximally stimulated by glutamate via NMDA receptors, but much less so when glutamate acts via quisqualate receptors. However, once NMDA-mediated PKC transfer has been initiated, it is additive with input via quisqualate receptors since it produces delayed, cumulative increases in responses mediated by that receptor.

This cumulative effect (like that of the high-frequency stimulation inducing LTP) may depend upon an intracellular positive feedback such as the IP3-mediated increase in available calcium described by Harootunian et al. (loc. cit.). It is also clear that if, in this system, dendritic PKC is acting (as in the adrenal cells described by Rasmussen, 1989) as a calcium pump, the result at the synapse was not a negative feedback. For if it were, PKC inhibitors should have lengthened, not shortened, the duration of LTP, and intracellular injection of PKC should have had the opposite effect.

The results of Davies et al. favor the conclusion that dendritic PKC, if in fact it serves as a calcium pump, may act as a positive feedback at the synapse. By establishing a pro tem loop between the presynaptic input(s) and the particular dendritic sites involved, such feedback would amount to a short-term memory mechanism. That is, it would act as a kind of dynamic or RAM memory, linking a downstream effect (LTP) with its upstream sources.

In turn, this is one way LTP might lead to the response specificity or selective hysteresis required of memory mechanisms generally. To visualize this effect, look again at the panels in Fig. 4-1, substituting glutamate for angiotensin II, and adding an axon terminal as its source.

If the site at which DAG and PKC combine to pump out calcium is sufficiently close to the one representing glutamate input, the result may be to make more calcium available for re-uptake at that synapse, as opposed to others further away on the same unit. In this way, calcium stored in calcisomes in the downstream cell might act to drive site-specific responses initiated upstream.

Negative feedback limiting this driving effect may be provided by some glutamate receptors, since it has been found that agonists, including adenosine and glutamate, block the N-type high-voltage-activated calcium channel in CA1 and CA3 neurons (see Chernevskaya, et al., 1991, and accompanying note).

The fact that, following LTP, the augmenting effect on quisqualate-mediated responses is *delayed* suggests that in the interim, transfer of PKC to other dendritic loci may be occurring, the delay reflecting the time required for this additional transfer. In other words, in LTP the dendritic transfer of PKC may, in fact, be site-specific; but as in the study by Davies et al. (1989), it may carry over, after an appreciable lag time, to include other sites concurrently activated by way of other receptors.

Müller and Connor (1991) present evidence that "in living brain slices" stimulation of associational-commissural fibers resulted in calcium accumulations specific to certain dendritic spines on CA3 pyramids. This finding appears consistent with the model just outlined, leading to the conclusion that differentially high concentrations of PKC might be found in the same dendritic spines.

The net result would then be a mechanism that pro tem conserved the pattern of a group of converging inputs—it being in this way that LTP acts to bring a number of concomitant events to a focus in a given group of "dendritic compartments." That is, it mediates Hebbian convergence effects, which in turn will tend to be maximized by the nonspecific (reticular) components of "coactivation" described in chapter 5, part 3, below.

In Alkon's rabbits, PKC transfer to dendritic membrane was detectable three days after conditioning. Note that these data were obtained not from cell slices (where LTP persists, as above, on a scale of hours), but from whole animals. In Alkon's rabbits, circular interaction of the hippocampus with the cortex via the perforant path may consequently have acted to prolong LTP (and dendritic PKC transfer) in the hippocampus itself.

If the foregoing analysis is on the right track, it suggests that memories stored in "dendritic compartments" of CA1 cells may be short-term or comparatively fluid. That is, local concentrations in dendritic PKC will tend to change over time, in response to the shifting patterns of sensory and brainstem motivational inputs, and like those elsewhere in the nervous system (or in adrenal glomerular cells), may consequently be temporary. There is no evidence to suggest that lasting changes in postsynaptic membrane may also occur.

In other words, the additional morphological changes equivalent to fixation of a long-term memory may not occur in the hippocampus proper. In contrast to GAP-43, whose chief effects may be presynaptic, the concentration of growth factors such as NT-3 (Phillips et al., 1990), which have been found in a subset of hippocampal neurons (or similar factors such as BDNF; see next section, below), may act postsynaptically. That is, they may principally act upon the receptor membrane, not of hippocampal units but of the cortical units to which they project. As a result, data transiently represented in the hippocampal system may, by way of these efferents, become more stably represented in parts of the cortex.

For this reason, Milner's Henry and Squire's patient R.B., while they could make no further additions to LTM, also apparently showed no significant *loss* of memories acquired before lesioning. Likewise, a more recent study by Zola-Morgan and Squire (1990) demonstrated that in the monkey, memory functions of the hippocampus proper are "time-limited"—i.e., comparatively short-term.

Part 2. Summary of the Argument to This Point

Abstract. Summary of the hypothesis of memory organization to this point. Long-term memories as consisting of stable "core" elements whose other referents, being subject to further probabilistic change, may alter or drop out of the memories of which they were originally a part. Hence the peculiarities of LTM as we experience it. Postsynaptic (dendritic) changes as the proposed basis for cortical memory formation in general. Trophic factors such as NT-3 as figuring in the fixation of such dendritic changes, and thus in the formation of long-term memories in particular.

The hypothesis of (long-term declarative) memory functions developed to this point comes down to the following propositions:

(1) That at the unitary level, LTM formation may involve lasting site-specific changes in cell receptor membrane, and is hence postsynaptic. (By contrast, memories formed on Blakemore's principle, though also postsynaptic, are unfixated and hence subject to further probabilistic change.)

(2) That in the cortex, LTM formation may occur as a result of return input to the several areas of association cortex that have reciprocal connections with the FD-hippocampal system.

(3) That an early stage in this circular interaction, here termed "registration," is preparatory to, but may not necessarily result in, LTM formation.

(4) That when circular interaction between the FD-hippocampal system and these several cortical areas is sufficiently prolonged, LTM formation in the cortex may result. That is, the memory concerned becomes "consolidated."

(5) That consolidation may be a process by which the several "core" components of a nascent memory become linked, over pathways in which more labile referents of the same memory also form links. (Fig. 4-3, below.)

Thus the key features of a given LTM may be "fixated" by the site-specific changes mentioned in (1), whereas the referents of the same memory that link it into a working whole, may not be fixated, and accordingly not as reliable. These would involve cortical units or assemblies that continue to respond, on Blakemore's principle, to the shifting probabilities of day-to-day experience.

The consequent changes in the associative connections of these units or assemblies, occurring locally at one or more stages in the transcortical relay, will be reflected in the structure of LTM itself. Accumulating over time, such changes may falsify a given memory (by substitution of new details for old); may make it more schematic (due to loss-by-reassociation of original details or referents); or may, in the extreme, reduce its probability of retrieval to near zero.

That is, if enough of the pathways mediating retrieval become blocked or diverted as a result of associative changes in their connecting links, the corresponding memory may become "occluded" or strongly resistant to reactivation via the usual cortical routes. (It may still be accessible by way of affect-related retrieval, whose subcortico-cortical circuitry is discussed in chapter 7.)

The schematic shown in Fig. 4-3 illustrates, in highly simplified form, the way the system may work. It represents a one-modality cortex that (read from right to left) receives thalamic input to a single prime receiving area. For reasons explained later below, playback via a, b, and c reaches only as far as a (equivalent, e.g., to visual area 19). Playback is assumed to occur by way of return pathways fed into layer III of the interareal relay system at each of the stages shown.

Note that under this scheme, not all referents of a given memory may be directly in the path(s) connecting its "core" components. Some will have taken shape in parallel pathways converging on a common set of

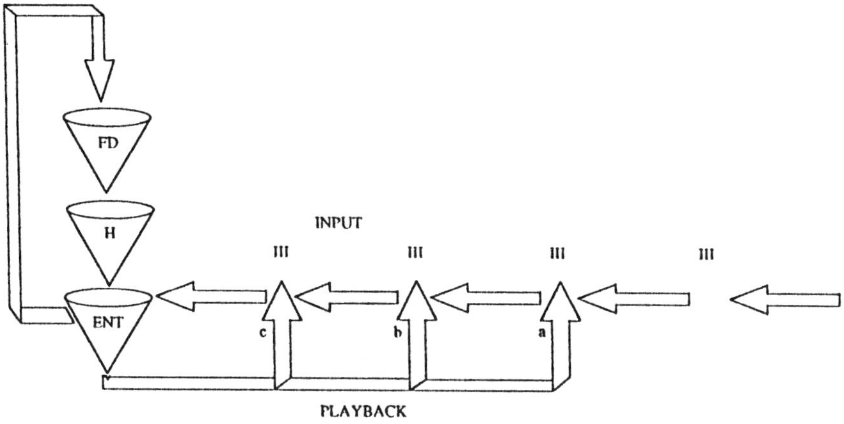

Figure 4-3. During consolidation, the "core" components or key descriptors of a given long-term memory are hypothesized to become fixated, e.g., by way of neurotrophic factors such as NT-3 (see text), as a result of playback from the hippocampal system and entorhinalis (here FD, H, ENT) to distal association cortex, e.g., of the fusiform or parahippocampal gyri. (In the schematic this would be represented by the return loop running ENT → c → III.) The other components of the same memory, though unfixated and therefore subject to subsequent probabilistic change, become linked to its core components by circular reinforcement—in the schematic, by positive feedback from ENT to a and b, forming a complete pathway (a → b → c → ENT).

It is postulated, but remains to be demonstrated, that the many return pathways from entorhinalis to sites in association cortex terminate chiefly in magnocellular lower III (as shown in the figure), thus feeding directly back into the interareal relay system. These are the pathways proposed here to figure in circular reinforcement.

The blank left between the two rightmost horizontal arrows is intended to symbolize those parasensory or proximal association areas that do not receive entorhinal afferents. Memories established in this cortex from the sensory input side are not directly retrievable into consciousness but, being elements in a linked system, can "prime" the retrieval of memories of which they are preconscious components.

assemblies. Others will develop, as it were laterally, by way of local associative processes running at right angles to the direction of throughput (i.e., out of the plane of the connections schematized in Fig. 4-3). Changes in such referents may falsify parts of a given memory without affecting its retrievability, this being one of the commoner consequences of the passage of time. (We tend to misremember details, forgetting some, making incorrect substitutions for others.)

Apropos of point (1) above, the most likely mechanism for bringing about lasting postsynaptic changes appears to be presynaptic release of one of the several known trophic factors found in the brain. That the hippocampus, in particular, is a source of such factors has been known for some time.

In vitro studies by Crutcher and Collins (1982) showed that hippocampal units produce growth factors capable of inducing neurite extension in chick ciliary ganglion units (parasympathetic) or in lumbar chain neurons (sympathetic). Kromer (1987) reported that after axotomy, the survival of units of the medial septal nucleus (which, along with the lateral septal nucleus, connects reciprocally with the hippocampus) was increased 350% by ventricular infusion of NGF, suggesting that it "can act as a neurotrophic factor for these neurons."

Ayer-Lievre et al. (1988) report that "cholinergic neurons of the septum-basal forebrain respond to exogenous NGF (nerve growth factor) by increasing the level of choline acetyltransferase" (mouse; rat). The hippocampus and FD "contain the highest levels of NGF mRNA in the brain." Levels in the cortex were somewhat lower. "No labelling was seen in granular cells of the cerebellum," a finding consistent with other reports of low concentrations of NGF mRNA in that structure.

Phillips et al. (1990) describe brain-derived neurotrophic factor (BDNF) and neurotrophin-3 (NT-3) as homologues of NGF (or as forming a three-membered "family"). They report that the distribution of BDNF mRNA in the rat forebrain significantly overlaps that reported by Ayer-Lievre et al. (loc. cit.).

The wide distribution of these factors and, in neocortex, the extensive cholinergic projections of the nucleus basalis, suggest that these trophic factors may figure in memory functions of the type proposed by Blakemore. ACh has also been suggested to act as a modulator of neuronal plasticity, e.g., during conditioned changes in the tuning characteristics of units in primary auditory cortex (Weinberger et al., 1990). Its role in "coactivation" is discussed below.

Phillips et al. report that "in contrast to BDNF, which was consistently expressed throughout the target areas of basal forebrain cholinergic neurons, NT-3 expression was limited"—chiefly to FD, CA2, and "the most medial region of CA1." Similar findings, particularly for BDNF, were found in human postmortem material. The distribution in the hippocampus resembled that for the rat. No significant hybridization occurred "in cingulate, temporal or parietal regions... in the amygdala or endopyriform cortex." There was, however, "weak to moderate labeling... in layer 2 of the pyriform cortex."

The hypothesis suggested by these findings is that the differential concentrations of NT-3 mRNA in the hippocampus reflect the presence of a growth factor that may have its chief effect on those cortical neurons reached by hippocampal efferents via the perforant path. The question then becomes: Do CA1 efferents end in entorhinalis or do some reach further into neocortex?

One wonders, also, if the presence of NT-3 mRNA in layer 2 of pyriform cortex may not be a means by which feedback from that cortex to mitral cells in the olfactory bulb acts to establish the "winner-take-all" response biases in those cells described in the computer simulation study by Ambros-Ingerson et al., (1990; cited in chapter 2 here). In other words, may LTM functions of the hippocampus be phylogenetically related to, or derived from, those mediating (learned) specific recognition in the olfactory system? (As odors depend upon the molecular structure of the odorant, olfactory memory needs to be similarly specific.)

In both cases, the distribution of NT-3 suggests that lasting memory formation may involve, not the cells that produce it, but chiefly those to which they project. What makes this hypothesis attractive is that it would explain why the hippocampus has been found to be indispensable to long-term memory formation in the cortex, e.g., in man.

What it would *not* explain is the "time-limited" memory functions reported for the hippocampus itself by Zola-Morgan and Squire (1990). For as NT-3 mRNA is also found in dentate granule cells, one might expect them to have an effect on CA3 and CA1 neurons like that produced by hippocampal efferents to cortical units. The same would apply to units reached by association fibers within the hippocampus itself. Why, then, does it not have an LTM-capability similar to that of (parts of) neocortex? (Might its greater phylogenetic age account for the difference?)

Except for that troublesome question, the data on NT-3 distribution, taken together with the cortical areas known to be reached by return fibers from entorhinalis, suggest the outlines of a system by means of which "core" components of long-term memories may be established in neocortex. Fig. 4-4 shows the areas-of-origin of afferents to, and efferents from, entorhinalis in the macaque.

It is clear that most of the pathways shown form loops, each arising from and terminating in approximately the same area. And if one compares them with, for example, the stages in the interareal relay of visual inputs shown here in Fig. 3-2, it is also clear that data fed back from the entorhinal area re-enter this relay at several stages, suggesting a connected feedforward system such as that shown in Fig. 4-3. The question is then whether re-entry at these points is chiefly by way of the relay

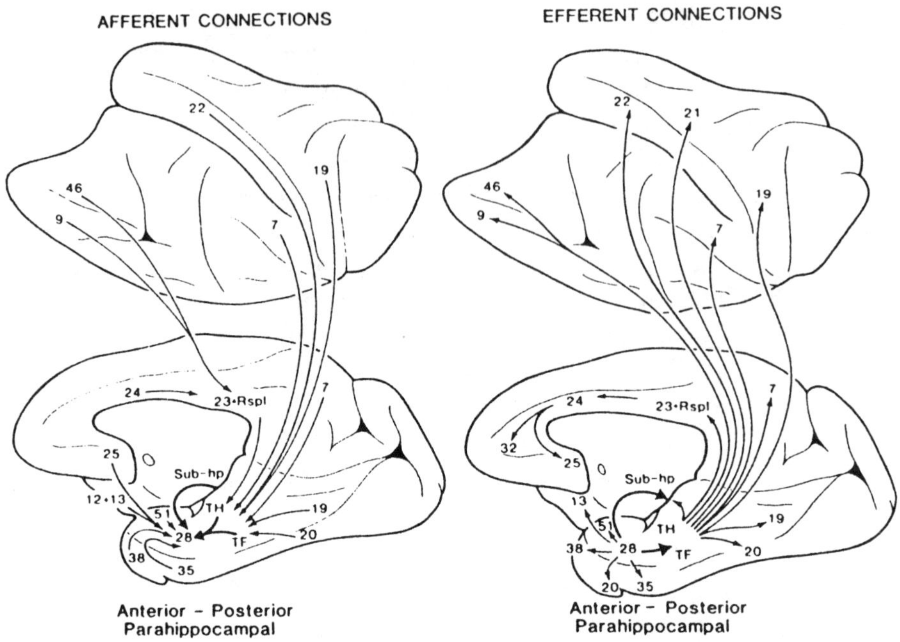

Figure 4-4. Schematic showing cortical areas in the macaque having afferent and efferent connections with entorhinal cortex. Reprinted with permission of *Trends in Neuroscience*, from Van Hoesen, GW (1982): *Trends in Neurosci* 5: 345–350.

system described by Pons et al. (1987). That is, do return fibers from the entorhinal area terminate principally in magnocellular lower layer III, as shown in Fig. 4-3? It is a reasonable supposition that that might be the case. How many (if any) CA1 fibers project beyond entorhinalis?

I raise these questions because they should be answerable by studies showing the laminar distribution of units responding to stimulation of these return pathways. Furthermore, latency data should indicate how many synapses are involved—how many (if any) of the areas shown in Fig. 4-4 may be reached monosynaptically by stimuli applied to perforant path fibers.

Finally, if activation of those fibers in fact entails release of a trophic factor such as NT-3, it should be possible to demonstrate its presence in their cortical terminals, as well as the cortical layers in which those terminals were chiefly concentrated. Such a set of results might, in turn, tell us which areas of association cortex contained target neurons subject to the long-term dendritic changes mentioned earlier (and which areas did not).

In the aggregate, units showing such changes would constitute the "core" components of long-term memories; and if represented initially in lower III (with concomitant changes in V, as described below), would form the more enduring links in the cortical memory network described in chapter 1.

As I have just tried to show, the model of LTM that emerges from this account has many of the features of long-term memory as we actually experience it. Consisting of patterns of generalized elements, the specificity of long-term memories consists in the pattern, not the elements; and many of the latter, because unfixated, may change, altering the pattern or dropping out of it. For this reason, long-term memory is not truly photographic, as Luria pointed out (1976). It is subject to errors in detail or to unaccountable "blocks," and does (as he said) tend to become sketchier or more schematic with the passage of time. The model proposed here would predict these properties.

It remains to consider points (3) through (5) above—the mechanics of registration and/or consolidation. Before doing so, I will take up a related question, concerning the mechanics of "coactivation" that Brown et al. (1988), among others, consider important in the induction of LTP and therefore in memory formation. The subsection that follows deals with the nonspecific component in coactivation, arising from forebrain projections of the reticular formation.

Part 3. The Role of the Reticular Formation in the Coactivation Leading to LTP

Abstract. The reticulocortical input system as supporting the "coactivation" leading to LTP. Essential neuroanatomic and physiological features of that system, including its cholinergic and noradrenergic subdivisions. Norepinephrine (NE) as potentiating both glutamate and acetylcholine (ACh) activation of cortical neurons. EEG and psychophysical correlates of NE-ACh activation of the cortex during rises in central arousal. Relation of these to LTP induction (via specific-nonspecific convergence, or "coactivation"), and subsequent increases in the probability of memory formation.

Following Hebb (1949), it has frequently been proposed that learning depends upon the convergence of inputs from different sources upon a given neuron. What has received less attention is the fact that such coactivation often includes, and may indeed require, convergence of nonspecific with one or more specific inputs to the same cortical unit(s).

In fact, LTM formation in single units may not necessarily depend upon Hebbian convergence, and may therefore not necessarily be "associational" in this conventional sense. The essential requirement may be that specific input(s) to a given unit include a nonspecific component that (as in tetanization) affects the whole cell. It is this component that is supplied by a sufficient rise in central activation and hence in reticulocortical input.

In this chapter, I will review data concerning the organization of the nonspecific projection system, the object being to show certain of the pathways which may figure in the coactivation of cortical units or assemblies, and thus in short- or long-term memory formation.

The reticular formation (RF) provides cortical input by a number of pathways and structures in addition to those long known—the thalamic nonspecific nuclei. By way of a cholinergic cell group in the midbrain, the RF projects not only to the thalamus, hypothalamus, and the preoptic area, but also to the important cholinergic system of the basal forebrain. The latter, in turn, projects to the hippocampus (via the diagonal band and medial septal nuclei), and widely, with some topographic ordering, to the neocortex (via the substantia innominata and nucleus basalis). This same basal forebrain system has connections with the ventral pallidum, forming part of the "limbic striatum" (cf. the report of Heimer et al., 1989, reviewed in chapter 8, below).

The ventral pallidum provides a feedback loop in this system, projecting to the subthalamic nucleus, dorsomedialis (DM), and parts of the amygdala. The latter, in turn, projects to the substantia innominata, a structure that "carries" and is also cross-linked with, the reciprocal ventral amygdalofugal path to the hypothalamus (Nauta and Feirtag, 1986, p. 243).

The reticular cell group having these rostral cholinergic projections lies ventral to the (also cholinergic) nucleus cuneiformis in the dorsal midbrain. N. cuneiformis projects to the tectum, thalamus, and globus pallidus (Scheibel, 1987). These two cell groups thus represent a degree of topographic ordering of reticular output at the midbrain level. (As mentioned below, the forebrain output of the dorsal and median raphe nuclei shows a similar ordering.)

Scheibel, for example, notes that "ascending reticular projections diverge at the junction of the mesencephalon and diencephalon. A dorsal leaf enters the thalamus and terminates among the cell masses of the thalamic nonspecific system." Whereas the ventral leaf accesses the cortex and "limbic striatum" by way of the basal forebrain, the dorsal leaf accesses the globus pallidus (i.e., part of the nonlimbic striatum), and reaches the cortex by way of the classical nonspecific system of the thalamus, which includes the midline and intralaminar nuclei and the netlike n. reticularis thalami.

The latter is a remarkable structure—"a sheet of gray matter that covers not only the rostral pole of the thalamus but also its ventral and lateral margins" (Nauta and Feirtag, 1986). It is GABA-rich and projects only inward, to the thalamus. Its inhibitory output is reportedly stimulated by "ascending thalamic volleys" (Scheibel, loc. cit.), whether by fibers of passage (collaterals) or by intrathalamic playback is unclear.

This arrangement appears to provide a normal waking-state control over thalamic output to the cortex. For Scheibel adds that some of the inhibitory control exerted by the thalamic reticular nucleus can itself be inhibited by "ascending impulses from mesencephalic reticular cells... presumably thereby allowing unimpeded thalamocortical communication." Undoubtedly, this is a mechanism that figures in the desynchronized "fast" EEG activity which typically accompanies central activation, or in the more transient cortical desynchronization that occurs during "novelty" or "orienting" responses.

In the auditory system, there are apparently RF offsets to the magnocellular medial geniculate nucleus, representing a direct intrusion of the nonspecific into the specific projection system. This nucleus is reported to have typically reticular (i.e., broad spectrum) response characteristics. It is activated not only by auditory but also by somatosensory or by "noxious" inputs, and projects non-tonotopically to all cortical auditory fields (Weinberger et al., 1990).

It appears to send projections to the centromedial amygdala (i.e., part of the "extended amygdala" of Heimer et al. discussed in chapter 8 below), and via that structure to the basal forebrain cholinergic projection system just mentioned (Weinberger et al., loc. cit.).

Rises in reticulocortical input tend to accompany stimuli having primary painful or fear-producing effects, including those acting upon, or arising from, the brainstem motivational systems concerned with feeding, safety, and sex. A low-level arousal or "orienting" response also tends to accompany any sufficiently sharp discontinuity in the pattern of incoming sensory inputs.

According to Scheibel, reticular neurons show rapid habituation to repetitive stimulation. For example, during stimulation of the sciatic nerve, the response of reticular units may drop to 10% of the initial value within 60 sec., an adaptation rate far higher than that, say, of cortical primary visual neurons, "which show response habituation patterns of no more than 5–15%" (Scheibel, 1987).

The RF can be visualized as a phylogenetically ancient core system that runs the length of the neuraxis (Bishop, 1958) and receives collaterals, at a number of levels, from the exteroceptive and interoceptive sensory systems. It is this basic arrangement that makes the RF a global monitor, whose labile responses uniquely reflect any changes in the physiological *status quo ante*. It is, par excellence, a system for the formation of temporary connections.

In the course of vertebrate evolution, this primitive system has differentiated so as to include subdivisions by way of which labile responses of the reticular core are amplified or extended in ways that support discrete special functions of phylogenetically newer parts of the central nervous system.

Three subdivisions of this kind are the noradrenergic locus coeruleus (LC), the serotonergic dorsal and median raphe nuclei and the dopaminergic ventral tegmental area (VTA). In these subsystems—the VTA in particular (see below)—the patterns of unitary response differ markedly from those described by Scheibel for reticular neurons. Along with input via Forel's tract to the thalamic nonspecific nuclei, they provide the major pathways by which the RF modulates activity in the neocortex.

Two of these pathways may be of particular importance to memory functions—the cholinergic of the basal forebrain and the noradrenergic of the locus coeruleus.

Before discussing in detail how these two may act synergistically, I should mention the dopaminergic projections of the mesencephalic ventral tegmentum (the VTA), and serotonergic projections of the raphe nuclei (which overlap the midbrain-hindbrain border. Nauta and Feirtag, loc. cit. p. 209).

The median raphe nn. project extensively to the limbic system and neocortex, the dorsal nuclei mainly to the striatum (Nauta and Feirtag, loc. cit.). Their function is not well understood, although studies of monopolar depression and of "violent" suicide suggest disturbance of a normal balance relationship, in which serotonin (5-HT) output is abnormally low and NE output correspondingly high (Cf. Fair, 1988, p. 179).

In neocortex, output of 5-HT neurons, like that of NE units, has been suggested to be diffuse or nonsynaptic. Two morphological types have been described by Kosofsky and Kowall (1989) for cortical 5-HT terminals; and by Vaney (1986), for serotonergic retinal amacrines.

Functions of the 5-HT system will not be discussed further here, nor will those of the dopaminergic VTA. The VTA although a differentiation of the main ascending RF, is one in which single unit firing patterns are quite unlike the labile, fast-adapting responses of reticular neurons reported by Scheibel (loc. cit.). VTA units are said to show tonic firing rates of 2–7 spikes/sec that are remarkably independent of the diurnal cycles affecting other systems, remarkably similar across species, and remarkably limited in their dynamic range during excitatory states or "bursting" (Jacobs, 1986).

Frontal cortex receives DA projections which are reportedly concentrated in the deeper layers, notably V (Bloom, 1981; Peterson, 1985). Many are reported to be inhibitory. However, concerning the laminar distribution of these terminals, and details of their morphology, see Goldman-Rakic et al. (1989, and appended note.)

During short- or long-term memory formation, two of these sources of nonspecific cortical input—the noradrenergic, of the LC, and the cholinergic, of the basal forebrain—may work together in ways I will now try to define, starting with the former.

Noradrenergic Cortical Input and Activation

Rises in central arousal tend in general to be accompanied by rises in LC output to the forebrain. (For a brief summary of inputs to the locus coeruleus, see the inset on the next page.)

It has long been known that in the cortex, "the major effect of activating this pathway is to produce an inhibition of spontaneous discharge" (Cooper et al., 1986. Similar inhibition of cingulate cortical units, induced by LC stimulation, was reported by Bloom, 1981).

Cooper et al. note that this is a "slow type of synaptic transaction, in which the hyperpolarizing response of the target cell is accompanied by increased membrane resistance." That this is a tonic phenomenon, affecting populations rather than particular units or assemblies, is suggested by the report of Fallon and Loughlin (1987) that norepinephrine (NE) may at least in part be released extracellularly rather than across synapses.

It now appears, however, that the effects of this noradrenergic input are more mixed and complex. Mouradian et al. (1988) report that in neocortex, "iontophoretically applied NE or activation of the coeruleo-cortical pathway is capable of facilitating neuronal responses to synaptic inputs and putative transmitters"—specifically, ACh and glutamate.

Using slices from rat somatosensory cortex, these authors recorded the unitary responses to iontophoretically applied ACh or glutamate occurring before, during and after iontophoresis of NE. Analysis of the data showed that during NE application, glutamate activation (in 54 cells) "routinely" showed potentiation of 94% versus control levels, and ACh activation (9 cells), potentiation of 67%. In 8 cells, NE evoked "robust excitatory

The LC reportedly contains only ~20–30,000 neurons per side in man (Nauta and Feirtag, 1986; Scheibel, 1987). In the rat (Aston-Jones et al., 1986) it receives minor inputs from the hypothalamic paraventricular nucleus and spinal gray near the central canal. Its two major inputs are from the medullary n. prepositus hypoglossi (which figures in the RF-mediation of lateral gaze control; Nauta, 1986), and from the nucleus paragigantocellularis.

The paragigantocellular area is of particular importance to the cortical arousal mediated by rostral projections of the LC. Aston-Jones et al. report that it receives projections from the interomedial lateral column of the cord, making it "a crossroads for circuitry pertaining to" the autonomic system. (Rises in LC output and in sympathetic activity tend to occur in parallel.) LC units "exhibit pronounced excitation following painful or polymodal non-noxious stimuli." In the medulla, pain responses are reported to involve the medial n. gigantocellularis (Casey and Jones, 1980).

According to Bloom (1981), β-adrenergic receptors are concentrated in layer II of rat frontal cortex. Lund (1981) states that the concentration of Gray Type 2 (i.e., inhibitory) neurons may be highest in layers II and III. Bloom reports that single unit studies have shown that LC stimulation resulted in inhibition of 50% of cingulate units tested.

Apropos of the reciprocal relation between NE and serotonin mentioned in the text, it is of some interest that in the hippocampal system NE and 5-HT terminals reportedly converge on a "restricted region"—the zona limitans of the dentate gyrus (Moore, 1977).

discharges in response to otherwise subthreshold iontophoretic doses of Glu."

The specific α-receptor antagonist phentolamine "routinely" blocked potentiation of threshold responses to glutamate. Activation of PKC by application of a phorbol ester mimicked the potentiating effect of NE on glutamate-evoked responses.

The authors add that "unlike noradrenergic augmentation of GABA-induced inhibition, which results from activation of the β-receptor linked cAMP system, these actions appear to be mediated by an α-adrenoceptor mechanism which may be linked to intracellular activation of protein kinase C."

These are extremely important findings, since they suggest that increased NE input from the locus coeruleus acts doubly to bring cortical activity to a focus during arousal. By way of GABA-ergic interneurons it produces a general tightening of inhibition. And at the same time, possibly by way of the α-1-receptor-mediated activation of PKC just mentioned, it potentiates responses to ACh and glutamate—two transmitters that play a major role in "coactivation." (Glutamate in addition, plays a major role in the residual facilitation leading on to consolidation, as described in part 4 of this chapter, below.)

Central activation is also likely to entrain simultaneous or immediately sequential rises in cholinergic input, originating in the mesencephalic reticular nuclei described above, and accessing the cortex by way of the substantia, innominata and n. basalis. (The nucleus basalis contains large multipolar neurons that show a branching pattern typical of the RF. Its topographic organization and subcortical connections are described by Tagliavini, 1987.) This cholinergic cortical input, being topographically organized, may be more phasically acting or area-specific than NE input from the locus coeruleus.

It amounts to a chiefly facilitatory second wave, converging with concurrent specific inputs to produce foci of reinforced ongoing activity. The reinforcement results from the double action of NE input, which produces increased background inhibition—essentially, a "contrast" effect—accompanied by potentiation of responses elicited by ACh, and subsequently by glutamate (as in the induction of LTP).

Note also that during arousal, the accompanying rise in reticular input to the thalamus may act, as described by Scheibel, to inhibit inhibitory functions of the reticular nucleus, thereby increasing thalamic outflow to the cortex. Concurrently, the increased inhibitory component due to NE input from the locus coeruleus will then, in effect, transfer reticular control of specific sensory inputs from the thalamic to the cortical level.

From the psychophysiological standpoint, this may be an extremely important mechanism, and one which is powerfully complemented by arousal- or fear-related responses of the amygdala (chapter 8, below). It implies that in situations which the organism interprets as requiring special attention, and responds to with rises in central activation, routine inhibitory management of sensory input at the thalamic level is temporarily set aside.

As a result, information flow from the thalamus to the cortex is increased. At the same time, ACh input (which shows some topographic organization), and NE input (which potentiates ACh responses while tightening cortical inhibition in general) will combine to increase the "processing" of currently or just-recently received sense data. (Interestingly enough, a study by Sara et al., 1988, shows that in the septal-hippocampal system, the *balance* of ACh to NE input is important to spatial learning in the rat. If one is reduced, the other must also be for learning to occur.)

The fact that NE potentiates glutamate-mediated responses even more strongly than it does those mediated by ACh (Mouradian et al., 1988) then suggests a clear sequence of steps leading to LTM formation. Initially, specific-plus-nonspecific input leads to "coactivation" in assemblies at certain cortical loci. In turn, via potentiated responses to glutamate, coactivation leads to LTP at certain of those same loci. Finally, if the accompanying response in the hippocampal system is sufficiently intense or prolonged, LTM formation involving circular interaction with some of those same cortical assemblies may result.

In an earlier study (Fair, 1988) I used this model of the sequential effects of NE and ACh input (during arousal) to account for some psychophysical findings reported by Weichselgartner and Sperling (1987). According to the model, rises in LC input to the cortex would be reflected in surface-negative potentials in the EEG—roughly, the N-100, or potentials such as the contingent negative variation (CNV) or the *Bereitschaftspotential* (preceding intended motor activity by 800 msec. Kristeva, 1984). These represented a noradrenergic or "leading-edge" inhibitory component of central arousal, a "clear" pulse. (Evidence for the facilitatory component of the NE input, or its potentiating effects on ACh or glutamate transmission, was not yet clear, and not included in that model.)

Surface negativity, in these slow potentials, corresponded to activity in NE terminals close to, or in, layer I. (Some but not all studies, show these as having a "strong tangential orientation" in layers I and VI (Bloom, 1981). Contra, see Routtenberg, 1978. There are, in any case, probably areal differences in this distribution.)

Subsequent positive-going potentials (e.g., the P-300) corresponded to activity in the middle cortical layers (e.g., IV and lower III), representing a facilitatory sequel ("reset"), arising from cholinergic projections of the basal forebrain. Some electroencephalographers have identified the P-300 or later positive-going potentials with cortical "processing." Desmedt (1981), in addition, suggested that the P-300 may be cholinergic. See also Decker and McGaugh (1988) on the probable involvement of cholinergic-noradrenergic interaction during learning.

The model just outlined showed a good fit with the time relations reported in the Weichselgartner and Sperling psychophysical study. This account was also consistent with the fact that scopolamine, a cholinergic antagonist, produces short-term memory (STM) deficits in normal controls resembling those shown by Alzheimer patients (Cole et al., 1985) in whom cholinergic forebrain projection systems are extensively damaged.

The reason that catecholamine blocking agents do not have similarly severe effects on short-term memory may be due to the fact that catecholamines have the contributory function described here. That is, they figure in alerting and attention, or as adjuncts to learning, but not constitutively in learning itself. (As proposed in chapter 8, the same is probably true of the amygdala.)

A role for norepinephrine (Stein, 1977) or epinephrine (McGaugh, 1985) in memory functions has often been suggested. The more recent findings of Decker and McGaugh (1988) are particularly relevant to this discussion and are summarized in the inset below.

Decker and McGaugh (1988) studied the effects of scopolamine, a muscarinic antagonist, on NE-depleted and on normal mice, using two learning paradigms—passive (inhibitory) avoidance and "place-training in a water maze."

Scopolamine, given i.p. ~20 minutes before training impaired retention in both tasks.

NE-depletion, down to an estimated 44% of control levels in the hippocampus "did not affect performance on either task and did not alter the effects of scopolamine." It did however, "impair retention of place-learning when mice were retested" at 16 days after initial learning. They note, in this connection, that aging rodents with deteriorating NE function, reportedly show normal learning but impaired retention.

The passive avoidance task is comparatively simple (CS signals "no-go"). Hence the inhibitory component in NE input—the focussing effect—was minimally called upon. This account would predict that the effect of scopolamine on retention would also have been less in the passive avoidance task than in place-training.

Learning in the latter situation *would* have required that the NE focussing effect be maintained in the cortex (and supplemented by NE input to the hippocampal system, thereby affecting responses of "place" neurons). For if this input were not maintained for sufficiently long at a sufficiently high level, the result would be to cut "registration" short, thereby precluding the next stage, or consolidation. See below, part 4 of this chapter.

It has been stated, for instance, that projections of the locus coeruleus figure in context-dependent or cue-facilitated recall (Sara, 1985; Deweer and Sara, 1985). Wahler and Hahn (1986) describe such cues as "those antecedent events that 'set the stage' for the occurrence of previously reinforced behaviors." That is, such contextual cues "tend to increase the likelihood of responding, without demanding the behavior."

Because of its sensitization-like (i.e., diffuse downstream) effects, cholinergic cortical input, as well as some of the downstream effects of LTP, would be likely to produce such contextual inclusions in memory. (That is, the role of NE in these effects, though important, would be secondary.) They would amount to recalled peripheral features (of significant events, persons, etc.), some of which would have a useful anticipatory function, and some of which would simply be irrelevant detail. Consolidation of long-term memories often includes such detail.

To summarize: the model just outlined proposes that two forms of nonspecific cortical input, arising in the noradrenergic locus coeruleus and in the cholinergic basal forebrain, provide a first-stage mechanism which may be of major importance in cortical memory formation. The cholinergic component is chiefly facilitatory, shows some topographic organization, and acts additively with excitatory activity already in progress.

The noradrenergic component includes both inhibitory and facilitatory effects. The former acts, via α-1 receptors, to increase general inhibition (by way of GABA interneurons). The facilitatory component acts additively with ACh input to increase activity in units or assemblies most active at the time—which usually means those immediately or most recently in receipt of specific inputs from the periphery. The result is to highlight those inputs both directly and by a concomitant inhibitory reduction of other competing cortical activity. In this way, the stage is set for memory formation as just described. (See Appendix III.)

It is unnecessary to add that this model of the mechanics of coactivation is greatly simplified. For example, it takes no account of the role that the midline and intralaminar nuclei may play in the process. (Single unit studies in these nuclei suggest distinct populations that respond differentially to stimulation of "aversive" or "reward" areas in the midbrain RF or medial forebrain bundle, respectively. Keene, 1975.)

Release of neuropeptides may also frequently accompany rises in central activation (as in the model proposed for ventral horn units by Hökfelt, et al., 1985). As noted by Krieger (1985), central nervous concentrations of neuropeptides are several orders of magnitude lower than those of the standard neurotransmitters, suggesting that they form a booster system, or (to quote Crawley, 1990) that they mediate "unusual neuronal responses to extreme physiological conditions."

Learning—in particular the protracted process of LTM formation—may consequently entail the local synergic action of a considerable assortment of neuropeptides, many of whose effects are only beginning to be understood. They include CCK, VIP (Fig. 1-1 above), somatostatin, vasopressin, neurotensin, calcitonin gene-related peptide (CGRP), and galanin (GAL), which is colocalized with NE in the locus coeruleus, with 5-HT in raphe units, and with ACh in the nucleus basalis and septohippocampal system, in several species including man (Crawley, 1990).

I have omitted this topic because, though many neuropeptides besides those mentioned may have a role in memory formation, we still lack definitive evidence as to what that role is. Partly that is because we have not yet defined what memory itself is—what structural changes underlie it, and what distinguishes known forms of memory from one another.

I might note, in this connection, that if NE favors "coactivation" via the dual action described here, one might begin to look for neuropeptides that could shift the balance of its inhibitory and facilitatory effects—a shift perhaps related to the role Crawley suggests for neuropeptides as modulators of "neuronal responses to extreme physiological conditions."

My aim in this section has been to show how two principal forms of reticulocortical input may act together to produce "coactivation" and thereby the conditions conducive to LTM formation. Much confusion has surrounded the question of what roles these cholinergic and noradrenergic inputs, and "event-related" cortical potentials, play in memory functions.

Squire (1987, p. 204) says that "in view of the finding that the (P-300) does not reflect medial temporal cortex function, the processes indexed by the (P-300) wave would appear to be part of early-stage attentional and short-term memory processes." This, I believe, is correct. Involvement of medial temporal cortex probably does occur, but at longer latency. (However, see de Toledo-Morrell et al., 1991, and accompanying note.)

The N-100 and P-300 waves correspond to the sequential rises in noradrenergic and cholinergic input to the cortex that tend to accompany rises in reticular activity or "arousal." They are therefore components of "novelty" responses, and as such, likely to lead to hippocampal registration of concurrently incoming sense data. The result is then the transient enhancement of short-term memory called the "primacy" effect.

These, in short, are starting conditions, necessary but not sufficient. Only if the arousal in which they originate is sufficiently intense or prolonged, will it lead to LTP-induction and circular reinforcement, and thereby to long-term memory formation.

Part 4. Registration as the Initial Phase of LTM Formation and Consolidation as Its Final Phase

Abstract. Registration as the first phase of hippocampal-cortical playback, and consolidation as its more prolonged or final phase. Registration acts to extend short-term memory (STM) without fixation (of "core" memory components). Such fixation is proposed to occur during consolidation, and to involve site-specific changes in postsynaptic membrane, mediated, e.g., by way of one or more immediate early genes (IEGs) and their mRNAs.

Because LTM formation tends to involve data that are motivationally or emotionally significant, it entails accompanying rises in reticulocortical input, or the "coactivation" described in the previous chapter. And because projections from the hippocampus and entorhinalis do not reach all of posterior sensory cortex, some memory data represented in that cortex are excluded from consolidation. These are the data that support the preconscious psychological phenomenon known as "priming."

At the start of part 2 of this chapter, it was proposed that in posterior cortex, LTM formation occurred as a result of playback or circular interaction over pathways reciprocally connecting the FD, hippocampus, and entorhinal cortex with the several association areas shown in Fig. 4-4.

It was further proposed that an early stage in this circular interaction or playback resulted in "registration." If playback ceased at this stage, the half-life of the corresponding cortical short-term memories would be extended, but the memories would not go on to become "consolidated" or part of LTM.

Registration, in other words, sets up the initial conditions for LTM formation, and figures in extensions of short-term memory such as those in the primacy effect (described at the end of this section). For consolidation to occur, playback must be maintained for a sufficient length of time or (over shorter intervals) at a sufficient level of intensity (essentially a firing-rate factor).

That this tradeoff relationship between time and intensity factors actually exists is suggested by the fact that long-term memories established at relatively low levels of central activation—memories acquired, say, through study—tend to require considerable repetition or practice. Memories acquired under the stress, e.g., of extreme fear or anger, may form after a single "take" and may last a lifetime.

In animal conditioning studies there is also, within limits, a relationship between stress and speed of learning—a relationship that can take the form of an inverted U function if the level of stress becomes too high and/or if the response to be learned is made too difficult. (Fair, 1988. Denenberg, 1967.)

The reason consolidation depends upon prolonged or, alternatively, unusually intense playback is presumed to be due to the fact that, in cortical units, it entails site-selective changes in postsynaptic membrane, mediated by presynaptic release of neurotrophic factors such as BDNF or NT-3 (part 1, this chapter). Stimulation-induced synthesis of such factors in the presynaptic neuron, and the dendritic changes they then mediate postsynaptically, may both involve immediate early genes (IEGs) and their corresponding mRNAs, and thus occur on a more extended time scale than the neuronal events in which they originate.

Units or assemblies in which such postsynaptic changes occur are presumed to be those in which the "core" components—the fixated elements—of long-term memories are established.

The process as a whole is then proposed to involve two phases. The first phase depends upon coactivation of cortical units by converging specific and nonspecific input (as described in the preceding section) and can occur on a scale of seconds to minutes. Depending upon the level of general arousal, and the affective or motivational state of the organism at the time, playback from the FD-hippocampal system will result in "registration"—extension of cortical short-term memory. (For reasons given below, these extensions are most likely to involve the first items in a new stimulus train. Note also that "coactivation" and "registration," having similar brainstem sources, are likely to run in parallel.)

For the second phase, or consolidation, to occur will then depend upon maintenance of "circular reinforcement" on a scale ranging from minutes to days. During this phase, playback from the hippocampal system and entorhinalis, over the pathways shown in Fig. 4-4, acts as a selective feedforward (Fig. 4-3), prolonging the foci of activity initially established in those pathways during the first or coactivation-plus-registration phase. Playback thus provides a means by which coactivation itself can be extended, and registration can lead on to LTM formation.

This may be an important mechanism and one that illustrates the use of loops in the central nervous processing of sense data. (Another is the fronto-posterior loop that figures in the generation of spoken or written language. See chapter 6, part 2.) Consequently, in man, when the hippocampal-cortical loop is cut bilaterally in the hippocampus, no further additions to long-term memory can be made.

In effect, coactivation sets up a cortical marker system such that, during playback, hippocampal-entorhinal efferents "find" the correct units or assemblies for inclusion in a nascent LTM. The short-term effect of playback is registration—the tagging of items for possible inclusion. The longer-term effect is consolidation, during which some of the tagged items in these recursive pathways become fixated as "core" components of a given long-term memory.

If this is, indeed, the way consolidation works, it would account for the fact (1) that LTM formation requires an intact hippocampus; (2) that it occurs by "backward" projection or "irradiation of the trace" (Flexner et al., 1963); (3) that it can be blocked by a protein synthesis inhibitor such as puromycin (ibid.); and (4), that it is time-dependent, usually requiring some minimum interval to go to completion. (Data illustrating this time-dependence are cited in chapter 3.)

In contrast to the mechanisms of postsynaptic "fixation" possibly mediated by hippocampal playback to one or more of the areas shown in Fig. 4-4, those maintaining residual facilitation in the cortical units reached during playback may be of the sensitization or presynaptic type, e.g., involving ACh initially and (in some areas or assemblies) LTP subsequently. These act to maintain the "marker system" that guides playback. (See, in this connection, the LTP reportedly induced in layer III of cat motor cortex by tetanic stimulation of precentral cortex. This study, by Iriki et al., 1989, is reviewed in chapter 1 above.)

The extent to which the whole process (of registration, and subsequently of consolidation) goes to completion, will be quite directly related to the adaptive significance of the initiating inputs. The motivational or affective weighting given those inputs will be critical in determining whether they shall or shall not be selected for registration; or if registered, will then result in playback sufficiently prolonged or intense to cause their consolidation as long-term memories.

(Note that in this process, components such as LTP or the ACh-mediated coactivation described in the preceding section, being presynaptic, are likely to have sensitization-like or diffuse downstream effects. As a result, long-term memories themselves often include fringe irrelevancies and, because episodically formed, huge lacunae as well. So, when recalling some threatening event or person in our experience, we wonder why it sticks in our minds that he was wearing a bow tie or had his hair cut in a certain way, or that it was snowing outside. We may also *not* remember what preceded or followed this incident or exactly when it occurred. Both the inclusions and the lacunae are consequences, simply, of the peculiar way our long-term memory functions are organized.)

Note also the pivotal role of entorhinal cortex in this system, and the rather selective input which it reportedly receives via uncollateralized fibers from the ventral tegmentum. (Fallon and Loughlin, 1987. By contrast, frontal association areas receive collateralized input from the VTA and also from the pars compacta of the substantia nigra). These dopaminergic projections, referred to in chapter 3 above, may serve as a powerful gating mechanism, since they terminate in a major interface between the neocortical and hippocampal-FD systems.

The foregoing is not meant to imply that these projections are the only gating mechanism affecting the entorhinal-hippocampal system, merely that they are an important one, reflecting affective and drive states arising at the ventral tegmental level. See, for example, the cortico-subcortical circuits involving the mammillary bodies, reviewed in chapter 7 below.

The presence in the hippocampus of Type 1, as well as of the more weakly responsive Type 2, corticosteroid receptors, suggests that stressful or nociceptive inputs may also have a gating function with respect to LTM formation. In this case, the gating action of the DA system may not be the principal one. (See, in this connection, the study by Antelman et al., 1988, casting serious doubt on the many reports that prefrontal DA neurons figure significantly in responses to immobilization or other forms of stress, such as "mild" footshock.)

The importance of central activation (and hence of converging specific and nonspecific inputs) in memory formation is implied by the many studies showing that administration of strong central stimulants such as strychnine or metrazol immediately following training demonstrably improves retention of the CR in experimental animals (McGaugh, 1985). This would correspond, crudely, to the intensity or firing-rate factor mentioned above. Moreover, rises in the output level of the RAS tend to be accompanied by rises in circulating levels of ACTH and corticosteroids. As noted in chapter 3, the hippocampus appears to be a "unique hormone target area" in the brain (McEwen et al., 1987).

In adrenalectomized rats, synapsin levels were raised in the hippocampus, but not elsewhere, following corticosterone administration (McEwen et al., 1987), a result implying that the hippocampus, unlike other brain areas, contains Type 1 corticosteroid receptors. A mechanism of this kind, by which increases in synaptic efficiency were made to parallel increases in certain types of adaptive demand on the organism, would illustrate further the relation between stress and the tendency to long-term memory formation.

The fiber systems responsible for hippocampal-cortical playback extend possibly as far as parasensory areas such as 19 (e.g., as shown in Fig. 4-4, for the macaque) but if that far, probably no farther. Graybiel and Berson (1981) note an anatomic point that is perhaps significant in this connection—namely, that 17–18 and the areas of far-distal association cortex that also receive visual input are those parts of the system that are least connected with each other.

The fractionation of visual data in 17–18 reported by Livingstone and Hubel (1988) suggests that incorporation of data in that state into LTM might be inefficient, since to record that amount of detail would represent an additional amount of central nervous work, with a very small return in added adaptive value.

Berlucchi and Sprague (1981) found that retention of a complex visual discrimination in the cat survived ablation of 17–18, but not of 19. Area 19 was the first visual area from which early experimenters were able, upon stimulation, to elicit complete images in human subjects (as opposed to "sparks" or "flashes" from 17–18; Fulton, 1951).

Finally, if long-term declarative memories were to involve the same cortex as figured in the primary stages of sense reception—for instance, 17 and 18 in the visual system—they might contaminate immediate perception with data derived from memory to the point that the distinction between things remembered and things perceived might break down.

If, on the other hand, LTM were to extend only part way through the cortical memory gradient, the process would be more efficient, and the specificity of particular recollections would depend, not upon fine detail, but upon the particular patterns formed of their (somewhat generalized) components. This account appears consistent with some of the known features of human recall, one of which is described in an "experiment" reported by Sartre (1948; see accompanying note).

It also suggests that the psychological phenomenon of "priming" may involve memories laid down in those areas of parasensory cortex that are not directly part of LTM and therefore not reached by the return pathways mediating retrieval. Those memory data (which remain indirectly accessible to amnesics as well as to normal subjects) are preconscious in the sense that they can prompt conscious activity (e.g., recognition or identification processes) without themselves becoming conscious.

That they *are* memory data is implied by the report of Squire (1987; p. 156) that "priming in amnesia occurs for real words and not for pseudowords;" and by the fact that normal subjects, shown line drawings of geometrically "possible" and "impossible" objects, showed priming only to the former (Tulving and Schacter, 1990).

In conclusion, I might note that episodic rises in reticulocortical input, with some accompanying participation of the temporal-hippocampal system, occur as normal concomitants of a great deal of our day-to-day experience. The commonest form of this response, as mentioned above, is what Sokolov (1960) called the "orienting reflex"—a usually brief episode of arousal, with EEG desynchronization, triggered by any sufficiently abrupt discontinuity or change of pattern in sensory inputs.

A similar discontinuity arises whenever we start a new task, it being the accompanying transient rise in central activation that, in tests of serial learning, appears to produce the "primacy" effect. It is as though an alerting response, originating in the brainstem reticular formation, automatically raised input to the hippocampal system to the level needed to begin induction of LTP, and thereby registration.

That is, hippocampal output to the cortex briefly increases, giving the first items of a "new" train of sensory inputs a longer half-life in short-term memory than those that immediately follow. (And in most cases registration stops there. As in the majority of "orienting" or "novelty" responses, no LTM forms.)

The results in serial learning tests will usually also show a "recency" effect, essentially due to very short-term memory—a form of hysteresis equivalent to retention of what was last in the (cortical) sensory registers. (This account would predict that cholinergic antagonists would shorten or abolish recency effects, whereas noradrenergic antagonists would produce much smaller deficits, if any.)

These two forms of memory, one involving hippocampal registration (playback without consolidation) and the other not, appear to be experimentally separable.

Kesner and Novak (1982) found that rats trained in a three-way maze choice showed clear recency and primacy effects (for the respective maze routes), when tested at 20 seconds delay; at 10 minutes, the recency, but not the primacy effect had disappeared. After bilateral lesions in the dorsal hippocampus, the animals showed recency but no primacy effects at the shorter delay; at 10 minutes delay, they performed at chance level.

Squire (1988) states that the latter findings "accurately reproduce the dissociation of short-term from long-term memory found in H.M. and other amnesic patients." H.M. was "Henry," Milner's bilaterally hippocampectomized patient. In serial learning tests, he showed no primacy effect.

I should close by posing a question of considerable importance—the question of how far into association cortex the process of "fixation" may extend. Looking again at Figs. 4-3 and 4-4, we can see that playback, e.g., to areas such as visual 19, will act additively with Blakemore's principle to set up certain preferred through-put pathways. It may thus only be necessary for the fixation of what I have called "core" components to occur in more distal parts of the same pathways, since retrieval, set up in this way, will tend to find them.

In other words, the distal "core" components in these pathways will tend to be stable, whereas those leading *to* them will, as described above, be less so. That is, playback involves the whole of a given pathway, "fixation" only part.

That cortical memory functions are organized in this way is suggested by the fact that temporally, during perception, categorical recognition has been reported to precede specific recognition of a given sensory input (see Ambros-Ingerson et al., 1990). This, in turn, is consistent with the clinical finding that in the visual system, specific recognition can be blocked by bilateral damage to the lingual and fusiform gyri, whereas generic (visual) recognition, being apparently mediated by cortex lying closer to the prime receiving areas, remains intact (Damasio, 1985). It would also be consistent with the fact that old memories which have become resistant to ordinary conscious recall (by the "occlusion" described earlier), can sometimes be retrieved via affect-related inputs involving the brainstem circuits described in chapter 7.

Chapter 5
Laminar Organization

Part 1. Memory as a Function of Magnocellular Arrays. Learning on Blakemore's Principle

Abstract. Developmental learning versus learning in the mature organism. Memory as in general a function of magnocellular output arrays; in neocortex, as forming a bilayered system, chiefly involving layers lower III and V. Evidence in support of this conclusion, and of Blakemore's principle that single neurons respond preferentially to inputs they have "seen" most often. Generic memory as a function chiefly of newer cortex, and one that permits the organism to construct an inner world matching its experience. Evidence that motor learning may differentially involve pyramids of layer V, and be promoted by induction of LTP in neurons of overlying layer III.

Learning is usually divided into two basic types. The first—developmental learning—occurs early in the life of the organism, runs on species-specific timetables, and once established, tends to be fixed. Learning of the second type includes the several forms of memory discussed in this monograph. Its distinguishing feature is that it is not time-limited. It may occur (although at different rates) throughout adult life, and may, in addition, never become wholly fixed.

What both types have in common is that they result from changes in connectivity patterns, and therefore involve some of the same biochemical mechanisms. In mature learning, the change may consist of no more than an increase in the efficiency of one pathway at the expense of others subserving the same function(s). Developmental learning, particularly in its earliest stages, involves more radical changes.

These include a pruning away of terminal arborizations, or a reversal of early excessive synaptogenesis (Rakic et al., 1986), it being in this way that, for example, inputs from the ipsi- and contralateral eyes become segregated in the geniculocortical system. Note, however, that this process has an important feature in common with generic memory formation. It is probabilistic, with the result that it can be manipulated so as to produce lasting anomalies in vision (Spinelli et al., 1972). It prepares the growing organism to see what it is most likely to see, if in fact its early and later visual environments are statistically similar.

In addition to changes in the distribution and density of axon terminals, there appear to be early changes at the synapse, equivalent to differential increases in efficiency corresponding to the de facto probability (frequency) of the related inputs. It is presumably these differential changes that set up the receptive fields of neurons in primary visual cortex.

That this is an early, and soon terminated, memory-forming process involving parvocellular assemblies of layers II–IV is suggested by the recent report of Fox et al. (1991). They showed that in 3-week-old kittens, as compared to adult animals, the concentration of NMDA receptors in layer IV is high. At 6 weeks, this difference has lessened but not disappeared. These kittens were the controls. An experimental group that had been light deprived since birth, showed a significantly higher concentration of NMDA receptors in layer IV at age 6 weeks.

I would interpret these results as showing the effects of delay in the process of developmental learning in the chief receiving assemblies in primary visual cortex. In adult cats, that process had long since gone to completion, and the concentration of NMDA receptors in layer IV was consequently minimal. To the same extent, that part of the visual system had become "hard-wired."

Apropos of the memory functions proposed here for lower layer III, it might be noted that in adult cats, the concentration of NMDA receptors had *increased* from the levels found at 3–6 weeks, and was then highest for all layers sampled.

What this account suggests is that memory formation, understood as the experiential establishment of preferred connections, occurs earliest and most radically, in prime receiving cortex, and is there most resistant to subsequent change. In contrast, learning in outlying cortical areas may involve more limited changes, e.g., in axon terminals or in parts of postsynaptic membrane, and may (as in LTM) become only partially fixed or "hard-wired." It is significant, in this connection, that cortex of this type—the distal association areas once known as Flechsig's "terminal zones"—is the last to myelinate, the process, in man, reportedly continuing into the fourth decade.

We may, in other words, be talking about a plasticity gradient, such that in posterior cortex, the prime receiving areas are at maturity the least subject to further change, while those at the greatest synaptic distance from them are the most. The advantage of this arrangement is that it guarantees the stability of primary perception, while permitting some flexibility in the way these primary data are recombined and retained in memory.

At more distal ends of the system, playback from the hippocampus then provides for the selective stabilization of some of these throughput data as "core" components of LTM. The advantage of this arrangement is that it permits long-term recall of specific persons, objects, or events, but in a way that prevents the system as a whole from rigidifying to the extent that primary visual or auditory responses seem to do.

Its disadvantage is that, since long-term memories include many referents that are subject to further probabilistic change, the memories themselves tend to become increasingly unreliable with time. LTM, one might say, has proved just reliable enough to serve its adaptive purposes, and has.thus persisted, without radical improvement, in higher vertebrates including ourselves.

The Two-Tiered Cortical Memory System

In a paper written some years ago (Fair, 1965) I proposed that in six-layered cortex, memory formation might largely have come to be a function of magnocellular output assemblies in layers V and lower III. In chapters 1 and 2 here, I suggested that these arrays, which receive input from local parvocellular "processing" assemblies, as well as from other areas or from the thalamus, undergo threshold changes, such that the sum of inputs capable of driving them is held within some average upper limit.

What this means, over time, is that in proportion as a unit has repeatedly "seen" a particular pattern of input, the probability of its response to that pattern will go up, while that of its responses to other input patterns will go down. In proportion as the efficiency of some synapses on its receptor membrane increases through use, there will be a homeostatic decrease in the efficiency of others less used.

The studies of Kaas et al. (1981; 1990) imply such a dual mechanism since they show that partially deafferented neurons in primary somatosensory or visual cortex develop atypical responses, as if by a lowering of thesholds to previously available but "ignored" inputs. As such threshold changes are proposed to involve postsynaptic membrane, possibly including site-specific changes in receptor density as well as in overall dendritic morphology, they might be blocked by protein synthesis inhibitors, which in turn might block the reafferentation described by Kaas et al.

If, in fact, it is by this dual threshold adjustment mechanism that neurons come to respond differentially to inputs they have "seen" most often, the question is why that type of memory formation should chiefly have involved magnocellular output arrays. One reason may be that the convergence of input on units in those arrays is so great that it has resulted in a proportionate exaggeration of the homeostatic threshold changes just described.

Such units, in other words, are *relatively* memory forming. But in practice, this relative difference may be sufficiently large that we can regard mnemonic functions of parvocellular "processing" assemblies as negligible. One biochemical correlate of this difference would then be the differentially high concentrations of phosphate-activated glutaminase found in large pyramids of layers III and V (Akiyama et al., 1989; see the end of this chapter).

Over several stages in the cortical sensory relay, this memory system acts to put together highly particulate data (e.g., as represented in primary visual or auditory cortex) such that the most commonly encountered patterns are those most likely to be retained or later "recognized." The putting together results from the convergence of input upon units of large-celled arrays (chapter 2). The same process acts to exclude less or least probable patterns from retention, this result being essentially a by-product of the homeostatic threshold-adjustment mechanism just described.

Repeated across the cortex, the foregoing amounts to a process of successive invariance extraction. At each stage, the retained data represent a larger-scale putting together of most-probable features of day-to-day sensory experience, with increasingly cross-modal representation in areas such as Brodmann 39 (the angular gyrus) or finally in the hippocampus.

All of these representations are, in effect, reconstructed from sense data that in prime receiving areas are highly fractionated or pre-analyzed, notably as described for the visual system by Livingstone and Hubel (1988). As mentioned earlier, the advantage of this arrangement is the perceptual plasticity it confers. It enables the organism to put together in memory a set of recognitive structures corresponding to actual categories of its everyday experience.

Note that this categorical memory system develops by successive relay from newer parts of neocortex—the prime receiving and parasensory areas—to "proximal" and "distal" association cortex (Graybiel, 1974), forming the plasticity gradient mentioned at the start of this section.

Some categorical memories will reflect "learning dispositions" so strong, i.e., so closely tied to genetically guided developmental processes, that they will show little variation between individuals. These will tend to involve parasensory or proximal association cortex. An example, given in chapter 2, would be the sensitivity to relative motion (and the accordingly enlarged sensitivity to visual context) shown by MT (middle temporal) neurons in the owl monkey.

Memories of this type can afford to become "hard-wired" early—indeed had better do so—since they correspond to an environment in which things frequently move relative to one another as well as in relation to a stationary background.

In contrast to the stereotypy of MT responses, those obtained in inferotemporal cortex in the macaque (also discussed in chapter 2) appear to be far less predictable, suggesting that more distal association areas have been left comparatively "open." The function of cortex of this type is presumably to accommodate statistical trends in the organism's experience that cannot be anticipated by systems that lie closer to the primary input and tend to set earlier, during developmental learning

Because limbic influence is greatest in distal association cortex (Graybiel, 1974), what is selected for long-term recall there will tend to be heavily biased by concurrent motivational or affective states. The "fixation" of core components of LTM, e.g., in anterior, basal or mesial parts of this posterior cortex, will then further increase the unpredictability of unitary responses seen there, since such memories represent significant specific incidents (as opposed to the day-to-day invariants) of sensory experience. Hence the importance of the lingual and fusiform gyri to specific visual recognition in man. (Damasio, 1985. See Fig. 3-1 above.)

As illustrated in chapter 2, by the data of Gilbert and Wiesel (1981), the generalizing function of large-celled arrays can be seen to begin in prime receiving areas. In single units, such memory formation is a statistical resultant of sensory inputs and of the interaction of these with small-celled "processing" assemblies in the same column(s). Such processing will depend heavily upon the foci produced by inhibitory interneurons. Some of these are apparently intrinsic to small-celled assemblies, while others have basket-like terminals that envelop the somata of cortical pyramids (Lund, 1981), or units of the final output assemblies.

Colonnier (1981) notes that cells with these basket-like endings "are probably linked together by gap junctions, suggesting that they would be electrotonically coupled. A focal excitation of a group of these cells might thus result in lateral recruitment of other similar inhibitory cells . . . yielding a field of inhibition wider than the excitatory zone of input. This could be one of the mechanisms of the inhibitory surrounds of cortical columns."

The function of inhibition, as noted in chapter 4, is to bring local cortical activity to sharp foci, differentiating one set of processes from another or others. (Cajal seems to have inferred this principle before the physiology of it had begun to be known or understood. He suggested that the "delicacy of function" of the neocortex might depend upon the ratio of Golgi Type 2 neurons—small short-axon or axonless cells—to the larger pyramids. See Lorente de Nó, 1951.)

It is significant that in prime receiving cortex, granular layer IV is comparatively well developed but that thalamic afferents, besides having their main endings there, also have terminals that "extend well upward among the large deeply situated pyramidal cells of layer III" (Jones, 1981). Jones adds that outside of the prime receiving areas, or in about 80% of the cortex as a whole, input fibers "avoid layer IV and terminate mainly in the deep part of layer III."

En route from sensory to parasensory and finally to distal association cortex, layer IV progressively thins, and input is more directly to lower III, forming the interareal relay system described by Pons et al. (1987). As I will now try to show, it is one that accesses the upper tier of a two-tiered memory system, consisting of magnocellular output arrays in layers lower III and V. As is clear from Fig. 1-1, much of the output of V is to subcortical systems or (in area 4) to the final motor pathways, though according to Jones (1981) and White (1989) some long interareal fibers also originate in layer V.

Cells having mnemonic functions are presumed to be those that show changes in receptor (postsynaptic) membrane causing them to respond with higher probability to most frequently "seen" inputs, and with proportionately lowered probability to less or least frequently seen inputs. This is learning on Blakemore's principle, the presumption here being that it involves a subset of cells in cortical columns generally—the large-celled pyramidal output arrays of lower III and V.

This is a testable hypothesis. For if, in cortical columns, the responses of some assemblies are largely unlearned and therefore chiefly dependent upon dynamic maintenance by the local facilitatory-inhibitory circuitry, iontophoresis of a $GABA_A$ receptor antagonist such as bicuculline should distinguish them from memory-forming units in the same columns. Blockade of GABA-mediated inhibition should radically disorganize the responses of these non-memory-forming or "processing" assemblies.

However if, as proposed here, memory-forming units have undergone changes selectively affecting parts of their receptor membrane, they should be much more resistant to the disorganization resulting from GABA blockade. That is, pyramidal neurons of layers V and lower III that receive direct as well as indirect thalamic input should continue to show normal responses, whole those of smaller-celled processing assemblies should be disrupted. If units of V and lower III are *not* memory forming, GABA blockade should disrupt responses in all layers or throughout the column.

Bloom (1981) reviews experimental data showing that iontophoresis of bicuculline in primary visual cortex of adult animals (cat) caused receptive-field organization to be disrupted, as if by regression to the earlier "aspecific" type of response (Cooper, 1981) seen in neonates. These changes appear chiefly to have involved layers IV, II, and upper III.

To quote Bloom: "All simple cells tested lost their directional sensitivity" suggesting that "directional sensitivity for these cells arises within the cortex, as a result of a GABA-mediated inhibition.... The type 1 complex cell behaved to bicuculline inhibition like the simple cells.... Directional sensitivity was eliminated; all such cells were encountered above layer IV.

"The type 2 complex cell was unaffected.... These cells were concentrated largely in layer V, but some more superficial neurons of this type were also seen." (The level was unspecified, but strong inference from related data would suggest lower layer III.)

"The type 3 complex cells also showed directional sensitivity which was unaffected by bicuculline, and were also concentrated in layer V" (Bloom, loc. cit., p. 361).

These results suggest that, in fact, neurons of V and lower III may be memory forming, while others in the same column(s), e.g., in II, upper III, and IV, may not be. Neurons of the magnocellular layers presumably continued to show normal responses because (a) they were in receipt of thalamic input and (b) their responses to the input had become dependent upon site-selective changes in their receptor membrane.

The fact that those responses persisted, in spite of the diffuse or disorganized activity produced elsewhere in the same column(s) by GABA blockade, carries an important implication. It suggests that as memory-forming units had developed preferred or learned responses, their probability of responding to other inputs (e.g., from diffusely responding cells nearby) had decreased accordingly.

In other words, as proposed here, learning in single units may involve two sets of changes in postsynaptic membrane, such that a lowering of excitatory thresholds at certain input sites also entails rises in threshold at others. These reciprocal changes not only favor response-specificity by exclusion of competing inputs or noise but may, as suggested above, have a homeostatic function as well, in stabilizing or reducing net output of the cell.

In an earlier developmental stage, the establishment of such membrane changes, over time, may well have depended upon inhibitory processing of thalamic input, e.g in II, upper III, and IV. Once established, however, such changes made the responses of units in magnocellular arrays, e.g., in lower III and V, automatically selective or "learned" and therefore either dependent upon thalamic input alone, or upon selection of preferred from diffuse inputs, as in the bicuculline experiment.

In the mature animals used in the studies by Kaas et al. (1981; 1990), partial cortical deafferentation may then have required that the process of developing preferred responses be begun all over again. The fact that it occurred suggests that, after all, Blakemore was right—cells do respond preferentially to inputs they've "seen" most often; and if deprived of those, will develop preferred responses to other inputs still available to them.

Friedberg (1990) reported an operant conditioning study involving a (motor withdrawal) response mediated by cells of the rat somatosensory "barrel" fields, the response being triggered by stimuli to the vibrissae. Two groups of experimental animals were subjected to NMDA-lesioning, which ablated layers I–III of a substantial area of this cortex. One group of animals was trained in the response prior to lesioning; the other group was trained at the end of the recovery period following lesioning.

The group trained prior to lesioning showed retention of the response approaching that of the controls. The group trained after lesioning showed significant deficits, or failure to learn the response.

Thus, survival of layers IV–VI, in the pretrained animals, was sufficient to mediate activation of the final motor output pathways involved in the response. The fact that that was not the case if I–III were ablated before training suggests the two-tiered arrangement proposed above. That is, in a given "column," memory formation may begin in magnocellular lower III and be followed by, or result in, memory formation at the level of V.

The latter represent memories governing outputs that are mainly, though not exclusively, cortico-subcortical (including outputs to the basal ganglia and pyramidal tract). The former, or memories formed at the level of lower III, mainly govern output to other parts of the cortex as described by Pons et al. (1987), and thus tend to make the posterior cortical memory system an interlocking one—for instance by way of associational connections established locally at the several stages in the interareal relay schematized in Fig 4-3.

(It is associational processes of this type that Squire (1987) has suggested may result in a kind of second-order consolidation. Such a mechanism would account for the fact that our grasp of a given subject and our recall of the pertinent facts appear to increase as a nonlinear function of the number of things we come to know about it.)

The foregoing data and concepts are consistent with the findings of Iriki et al. (1989) reviewed in chapter 1 above. In that study, tetanic stimulation of postcentral cortex (cat) resulted in LTP confined to units of layer III in precentral (motor) cortex. Tetanic stimulation crudely mimicked the effects of specific-nonspecific coactivation discussed here in chapter 4. The chief effect of tetanization was on units of layer III in motor cortex, the rostral next stage in the interareal relay system.

Under the hypothesis developed here, this result would represent a first stage in the formation of a motor memory. Iriki et al. note that interruption of connections between sensorimotor and motor cortex has been shown to retard the acquisition of conditioned motor responses, without affecting performance of responses already learned. (Cf. the Friedberg study reviewed above.) Under normal conditions, the second stage of motor memory formation (as Iriki et al. also suggest) would then involve units of layer V in those columns in which layer III units showed LTP.

In a continuation of the study by Iriki et al. (loc. cit.), Keller, Iriki, and Asanuma (1989) report that, following tetanic stimulation of post-central cortex, 14 out of 17 cells in cat motor cortex exhibiting LTP were pyramids. "Only 3 neurons were sparsely spiny, nonpyramidal cells." All were described as being in layers II or III of motor cortex. Apropos of their original finding that tetanization of postcentral cortex induced LTP in units of layer III in motor cortex, Iriki et al. conclude that this "associative LTP" may provide "a basis for the acquisition and retention of motor skills."

The important point here is that 82% of the cells in motor cortex showing LTP were pyramids.

In part 2 of this chapter, I will review biochemical and other data relevant to these processes.

Part 2. Biochemical and Physiological Data Supporting the Two-Tiered Memory Hypothesis

Abstract. Biochemical evidence supporting the bilayered (two-tiered) model of cortical memory functions. Maximal concentrations of phosphate-activated glutaminase in large pyramids of III and V. Glutamate receptor types and subtypes: functional implications of their distribution in neocortex and in the hippocampal CA1–CA3 fields. Intrinsic rhythmic activity in deep layer V pyramids (rat; in vitro). Its possible relation to coordination of nonlocal activity during memory formation; or to current phase-locking theories of central processing.

The present hypothesis would predict that under normal conditions, the convergence of specific and nonspecific input, e.g., upon units of layer III in motor cortex would approximate the conditions crudely set up by tetanization at a postcentral site. In other words, normal sensory inputs, when accompanied by a sufficient level of central arousal, may set up a train of sites across the cortex, at which "coactivation" occurs, initially having its chief effect on assemblies of layer III. The result is that, in certain pathways in the interareal relay system, the probability of memory formation at both levels—lower III and V—is substantially increased.

In line with this interpretation, Bindman and Murphy (1988) report that in slices of rat neocortex (from what area is not stated), test shocks (0.1 Hz) to the subcortical white matter induced LTP in units of layers III, V and VI. Bath-application of d-2-amino-5-phosphonovalerate (DAP) prior to stimulation caused failure of potentiation in layer V (as measured by excitatory postsynaptic potentials) over a period of "tens of minutes." DAP (an NMDA receptor antagonist) when applied after establishment of LTP in units of III, V, and VI, "reversibly reduced its magnitude, showing that NMDA-receptor-mediated activity was also involved in the *maintenance* of LTP in neocortex" (italics original).

The last finding is of interest since sustained activation of cortical units, by way of glutamate and NMDA receptors, is a probable correlate of the level(s) and duration of central activation needed to support memory formation. In other words, this is one of the factors that may figure in the time-dependency of consolidation, and might involve the positive feedback mediated by PKC suggested in chapter 4 (a feedback that in turn may be limited by glutamate blocking of the N-type high voltage-activated calcium channel reported by Chernevskaya et al., 1991).

The Bindman and Murphy study also implies that layer V units were differentially sensitive to pretest application of DAP. Such a result would follow from the present hypothesis and also be consistent with the results reported by Iriki et al. (loc. cit.). For if cortical memory formation is a two-stage process beginning at the level of lower III, then LTP (or other forms of residual facilitation, however mediated) will begin there also, and may spread to involve V. Thus if DAP, in these experiments, had a partial blocking effect on units of III, one would expect this effect to be greater or maximal in layer V.

Apropos of the foregoing, Akiyama et al., (1989) report that in human neocortex, postmortem immuno-histochemical studies show that phosphate-activated glutaminase (PAG, which catalyzes conversion of glutamine to glutamate) stains most intensely in the larger pyramids of layers III and V (including Betz cells), with "moderate to intense labeling" in fusiform cells of layer VI.

As noted in chapter 4 here, glutamate activation by way of NMDA receptors tends to be longer-lasting and perhaps slower in onset. Activation by way of receptors of the quisqualate type tends to be faster or more phasic. The present hypothesis would predict that glutamate receptors of the latter type would be associated with "processing" rather than mnemonic functions, and be distributed in the cortex and hippocampus accordingly. The following data are relevant to that point.

In two studies (Keinanen et al., 1990; Sommer et al., 1990), these authors report that "fast" glutamate-induced neurotransmission is mediated by "a family of four abundant AMPA receptors." (AMPA = α-amino-3-OH-5-Me-4-isoxazole propionic acid. Pharmacologically, AMPA = quisqualate > glutamate > kainate.)

In the second paper, they describe "the existence and function of a second molecular version of the four AMPA receptors, generated by alternative splicing." Denoting these receptors as A, B, C, and D, they report that the cDNAs encoding each involve a sequence of 115 base pairs that exists in one of two sequence versions. "This segment encodes 38 amino acid residues within a conserved receptor domain preceding the predicted fourth transmembrane region, and hence is probably located intracellularly."

Without going into the further details of this complex study, I will merely point out that the authors call these alternative forms "flip" and "flop" versions of each of the four receptors just mentioned (A, B, C, and D), and add that evidence indicates "that the flip receptors, as a class, behave differently from the flop class."

For example: "In assemblies of two receptor types, one partner can be dominant with respect to the fast desensitizing current component, whereas the other can determine the steady-state component." They add that "such dominance does not correlate with module type." The meaning of this statement is not entirely clear, since if "module type" refers to structures, and the balance of flip to flop variants differs (as they report) in different structures, dominance relations might be expected to differ accordingly.

What is of particular interest here is the regional distribution of these receptor variants in the rat brain. *"The cell specific nature of the alternative splice is particularly evident in the hippocampus"* (italics added). CA3 synthesizes only "flip" mRNA, and only for receptors A, B, and C. Both flip and flop mRNAs for all four receptors are synthesized in CA1. Furthermore "a switch in expression between the alternatively spliced mRNAs occurs precisely at the boundary between the CA1 and CA3 areas."

This presumably refers to CA2 that, though a small array compared to CA3 and CA1, is not a very sharp boundary. (See Lorente de Nó, 1934, Fig. 2.) The transition occurring in this region, however, has extremely interesting functional implications.

Equally interesting, in its implications for the differential distribution of cortical memory-forming and "processing" assemblies outlined in chapter 5, part 2, is the report that in rat neocortex, flip versions of the A, B, and C mRNAs show "high expression in layers II, III, and VI."

This is the same pattern (A, B, and C) found for CA3, in which we earlier concluded transaction rates might be high, because of the interaction of mossy fiber input with ongoing activity arising from the associational-commissural fiber tracts. In the cortex, II, upper III (and possibly VI) correspond to the layers that I concluded, on the basis of the bicuculline experiment, might be non-memory-forming or purely transactional (concerned with processing).

If the flip version of these mRNAs mediates high-transaction rates, via what the authors call "the fast desensitizing current component," its reported cortical distribution is consistent with the hypothesis of cortical memory organization here. One might also conjecture that, in CA1, the presence of flop as well as flip mRNAs for all four receptor types reflects its capacity for sustained activity, and hence its role in LTM formation in the cortex.

These findings show a suggestive fit with some of the structural arrangements and corresponding processes supposed here to be integral to memory functions. This is particularly the case since other data, reviewed above, indicate that LTP, mediated by (the slower, more tonically-acting) NMDA receptors, involves layers III and V, and that glutaminase concentrations are maximal in the larger pyramids of these layers. For V and magnocellular lower III are the two "tiers" of the cortical memory system proposed here.

The authors conclude: "The functional characteristics of the two alternative exons in the four AMPA receptors, and the cell-specific expression of both forms of the receptors suggest an important role for the splicing-directed channel modification. The exon switch may underlie adaptive changes in neurons such as plasticity" (Sommer et al., 1990).

Still other findings, whose functional implications are similar to these, have been presented by Masu et al. (1991). They report cloning and characterizing "the complementary DNA of a metabotropic glutamate receptor (mGluR) coupled to inositol phosphate/Ca^{2+} signal transduction." They state that this receptor "shows no sequence similarity to conventional G protein-coupled receptors and has a unique structure, with large hydrophilic sequences at both sides of seven putative membrane-spanning domains."

The authors isolated an mGluR cDNA clone from a rat cerebellum cDNA library. They then investigated its intracellular signal-transduction action by injecting *Xenopus* oocytes with the "mGluR transcript, synthesized in vitro." Finally, they used RNA blot analysis and *in situ* hybridization to determine the distribution of this receptor in the rat brain. The following is a summary of their findings and conclusions.

First, they note that "the large hydrophilic amino-terminal region" of the receptor "has some sequence similarity to the AMPA/kainate receptors" (the "flip" and "flop" versions of whose mRNAs we have just discussed). That is, this receptor, like some versions of those, may have fast phasic action with proportionate demands on cellular metabolism.

In line with that idea, they report that "the cloned mGluR activates a PTX (pertussis toxin) sensitive G protein, leading to (inositol 1,4,5-triphosphate) formation and intracellular Ca^{2+} mobilization" (hence its description as "metabotropic"). They add that this receptor "differs in agonist selectivity from the known ionotropic glutamate receptors and shares no antagonist with any other glutamate receptors."

By the methods mentioned, they then found that in the rat, the highest concentrations of mGluR are in the mitral and tufted cells of the olfactory bulb, and in cerebellar Purkinje cells. "Most" thalamic neurons express mGluR mRNA, with "prominent expression" in granule cells and CA4 (hilar) cells of the dentate gyrus. They add that "interestingly" its expression is greater in CA2 and CA3 than in CA1 by an estimated factor of 10.

The latter finding is of particular significance, since the hippocampal distribution of mGluR is similar or identical to that of the "flip" versions of (receptors) A, B, and C reported by Keinanen et al. (1990) and Sommer et al. (1990). Both may reflect the high transaction rates inferred here for CA3. (See the Afterword below, concerning high levels of tonic activity as in general characteristic of processing assemblies, and intermittent, phasic activity as characteristic of memory-forming or final output assemblies.)

Similarly high transaction rates may reasonably be inferred for the olfactory bulb, thalamus, and cerebellar Purkinje cells; and it is interesting (in view of their inhibitory functions) that long-term depression in parallel fiber Purkinje cell synapses can be evoked by quisqualate but not by NMDA (Masu et al., 1991). These authors add a point of considerable importance to the argument here.

They state that "it has recently been reported that LTP in mossy fiber-CA3 synapses as well as in the stratum radiatum-CA1 synapses, is suppressed by PTX (pertussis toxin) pretreatment of the hippocampus.... In the stratum radiatum synapses, *PTX has been suggested to act on presynaptic terminals that are mainly derived from pyramidal cells of CA3*" (italics added).

These findings, in short, begin to add biochemical support for the idea, advanced here, that both in the hippocampus and in the neocortex, there are distinct populations of "processing" units characterized by high transaction rates (and correspondingly high metabolic rates). These then act as feeder systems to output assemblies whose metabolic turnover rates are lower, and whose response patterns show greater hysteresis—i.e., memory-like functions.

In the cortex, the feeder systems or "processing" assemblies consist of layers II, upper III, and VI; in the hippocampal system, they consist of the dentate gyrus and CA3. The assemblies showing greater hysteresis, or retentive capacity, are lower III and V (such that V > lower III) in the cortex; and in the hippocampus, CA1.

Apropos of other data, cited earlier, suggesting that pyramids of cortical layers V and lower III may in fact play a key role in memory formation, Silva et al. (1991) report that in slices of rat somatosensory cortex, "deep" layer V pyramids showed a tendency to rhythmic firing.

Some of these were "intrinsically bursting cells" that "generated repetitive rhythmic bursts at basal interburst frequencies of 5 to 10 Hz.... Epochs of rhythmic firing (usually 1 to 5 sec, but sometimes up to 20 sec) could also be triggered by spontaneous or evoked postsynaptic potentials."

With membrane potential clamped at −60 to −65 mv, "a majority of neurons in deep layer 5 (59%, n = 146) displayed sustained, rhythmic 5 to 12 Hz patterns, of either single spikes or bursts of spikes.... A 4 msec intracellular current pulse elicited a train of single spikes that lasted about 4 sec and then stopped abruptly."

Note that these are frequencies in the theta and alpha ranges, theta being a hippocampal frequency often associated with learning in animals, and alpha a waking resting-state cortical rhythm, or tonic background upon which the faster beta rhythms then episodically supervene. If layer V pyramids are the base, so to speak, of the cortical memory system, they might be expected to have a tonic pace-making function of this kind. They might also be expected to respond to episodic rises in central arousal by showing high sustained single-unit frequencies, corresponding to their phasic function, e.g., during the coactivation leading to memory formation.

Data displayed in Silva et al.'s Fig. 1B show that at a baseline membrane potential of −61 mv, a hyperpolarizing current pulse "evoked a seven-burst epoch with a mean interburst frequency of 9.3 Hz. Each burst consisted of three spikes, firing at about 150 to 300 Hz."

With rising current intensities, the former frequency "increased to 35 Hz or more, yet firing showed little or no adaptation, even when stimuli were several seconds long. This characteristic was not found in nonrhythmic neurons, which generated only one spike or a brief burst at threshold and adapted during strong stimuli.... This rhythmic behavior was observed in recordings made from deep layer 5, but not from layers 2, 3, 4, and 6."

In cortical slices cut horizontally, only the slice that included layer V exhibited rhythmic firing. Biocytin injection indicated that the layer V cells that showed rhythmic bursting or rhythmic single-spiking cells were pyramids.

Silva et al. add a morphological point that has important implications for the hypothesis of memory functions proposed here. They report that these pyramids had somata with profuse basilar dendrites in layer V, and apical dendrites that branched repeatedly in I and upper II but "infrequently" in III and IV. This arrangement (if it holds across species or in higher mammals) would suggest that in a given column or vertical cortical segment, the primary input to V is from overlying "processing" assemblies, rather than directly from lower III.

(Note that layer II, besides being influenced by input reaching III, also receives input from layer II nearby, via the shorter interareal projection system. The resultants of these inputs are then presumably reflected in responses of layer V units in the same column.)

Thus, data received chiefly at the level of lower layer III, during interareal relay, may reach V intermixed with the results of processing in overlying layers of the same column. Therefore, at any point in this system, the memory contents of the two "tiers" will be related but probably not the same. And of the two, the memories retained in V will be of more immediate functional importance, since V and VI are chiefly responsible for local outflows (to subcortical structures or the cord).

Apropos of current phase-locking theories of central processing (Singer, 1990; Gray et al., 1989), the layer V system might provide a highly flexible pacemaker, capable of giving temporal coherence to spatially distributed cortical events. As a form of parallel integration of incoming data, it might also figure in the (temporal) integration of the outputs or behavior that result. For the same reason, memory functions, which represent another form of temporal integration reaching into the organism's near or distant past, tend also to be concentrated on the output sides of cortical assemblies—in lower III in the interareal relay system, and in V, where the more final cortical outflows arise.

Recall that layers V–VI are phylogenetically the oldest, and the first developmentally to come on line (Sarnat, 1987). The same sequence holds for single units. Axons (outputs) develop first; elaboration of dendritic trees and myelination come later (Sarnat, 1987). Myelination of association cortex is completed last, and in man, reportedly continues into the fourth decade.

In earlier work (Fair, 1988) I suggested that cortical memory functions might be two-tiered, with those of the upper (layer III) tier being the more fluid. Whether that last is the case remains to be seen; but the evidence reviewed here suggests that this dual memory system may in fact exist, and that in a given area, the contents of the two tiers, though similar, may not be the same.

Part 3. Summary of the Argument to This Point. Stages in LTM Formation

Abstract. Review: the two basic stages in LTM formation. Factors establishing the "marker system" followed during consolidation. How the latter can be disrupted by the sensitization-like effects of prolonged central arousal. Factors at the single-unit level that distinguish "procedural" from "declarative" learning. Data from Alkon suggesting that motor learning may be self-limiting. Serial processing, in the cortex, as figuring in learning; parallel processing as supporting efficient retrieval of the learned.

Long-term "declarative" memories as forming a unique class, phylogenetically descended from olfaction-related associations mediated by the *primordium hippocampi* in the adjacent dorsal pallium of amphibians.

To recapitulate: the model of long-term memory formation proposed here involves two basic stages or sets of processes. The first occurs during the rising phase of central arousal, when the convergence of specific and nonspecific excitatory inputs results in "coactivation." The inhibitory (noradrenergic) component of this input establishes the foci at which coactivation is likely to occur. At these, the induction of LTP in pyramids of layers III and V (chapter 5, part 2) is then likely to follow, its effect being to prolong the facilitation initially established by the ACh component of nonspecific input (chapter 4, part 2).

It is this compound mechanism that sets up the "markers" that determine the pathways to be followed subsequently, during circular interaction between the hippocampus and neocortex, thereby leading to consolidation, as described in chapter 4.

It is to be emphasized that this playback mechanism may also play a part in the type of probabilistic memory formation described by Blakemore (1974). That is, it may give a motivational or affective bias to that process, and hence to the reinforcement-through-use of the nonfixated postsynaptic changes proposed to underlie that type of memory. Because of the cortical distribution of return fiber tracts from the hippocampus shown in Fig. 4-4, this biasing effect will tend to be confined to the more limbic-related or distal parts of association cortex.

(See, in this connection, the difference, described in chapter 2, between the stereotypic visual responses shown by units of MT in the owl monkey, and the more unpredictable responses of units in inferotemporal cortex, found in the macaque.)

The second stage of LTM formation overlaps and outlasts the first or primary arousal stage. This is the stage in which hippocampal after-activity and circular interaction with the cortex may lead to the consolidation of long-term memories, in ways outlined in chapter 4. The function of LTP, in this system, is essentially to prolong the activity set up in certain pathways by coactivation (converging specific and nonspecific inputs) during stage 1. LTP is what permits the sequels of that stage to take full effect.

Because LTM formation is a complex time-dependent process, it tends to occur during stage 2, the declining phase of the affect- or drive-related arousal in which it originates. This is an aftermath stage during which consolidation may be propelled by the tendency to after-activity of the hippocampus itself. Thus, while a very high level of central activation may produce correspondingly high levels of hippocampal after-activity (and so result in the "one-take" memory formation mentioned in chapter 4) this initial activation, or stage 1, may need to fall back to normal levels within some outside time-limit, if consolidation is not to be disrupted in stage 2.

For if the central activation occurring in stage 1 is unusually prolonged and intense, its effect may be to disorganize the "marker system" that normally guides playback during consolidation. In effect, a high rate of nonspecific input sustained for sufficiently long, may result in coactivation at too many cortical sites. The result is then to open so many pathways for playback to follow that the whole process loses focus and fails. Too many "choices" of paths amounts to no choice.

Instead of consolidation, what occurs is the kind of generalized response that Kandel and Schwartz (1985) describe as typical of sensitization (chapter 4, part 1, above). The description is apt, since in this case, it is the predominance of presynaptic driving and its diffuse downstream effects that produces a sensitization-like result.

In a previous book (Fair, 1988, pp. 138–143) I reviewed evidence from studies by Pavlov, by Maier, and by Melzack, indicating that when animals were put into learning situations in which the incentive to learn was strong but the task to be learned very difficult or impossible, the result was a regression of this kind.

In the Pavlov study—the clearest of the three—the animal was presented with a discrimination task that was at or just beyond its powers of visual resolution. Its successes consequently failed to rise much above chance level, and consolidation of the correct response was held up or blocked.

As it was repeatedly confronted with this hopeless task, the animal's arousal level began to rise, resulting in the presynaptic driving described above. At this stage, its initial gains in learning went into reverse, and spread to affect its performance on other easier tasks it had previously learned. It underwent a behavioral sea change, beginning to bark when brought into the lab, and to "wriggle" or bite at its harness. The learning deficits and generalized aversive behavior persisted for months afterward, suggesting a type of long-term presynaptic learning or sensitization so pervasive that even retrieval of pre-established memories was disrupted.

What these results appear to illustrate is the principle suggested above: that the pre- and postsynaptic components of memory formation depend upon a balance relationship between the two. If the presynaptic component, due to central arousal, exceeds some time and/or intensity limit, it can result in severe learning impairment (due to the disruption of consolidation just described), accompanied by some degree of subcortical release and behavioral regression. Some of the clinical implications of this account are taken up in that earlier work (Fair, 1988, pp. 141–143).

In speaking of stages 1 and 2 in LTM formation, we are in fact talking about waveforms—the slope of the curve representing rising and declining phases of an episode of central arousal. It is significant in this connection that the hippocampus, unlike most other limbic or neocortical structures, contains Type 1 as well as Type 2 glucocorticoid receptors. During intense arousal, cortisol, acting via the relatively high affinity Type 1 receptor, may serve to prolong the after-activity to which the hippocampus, with its single layer of large pyramids and its relative paucity of inhibitory interneurons, is notably prone.

In short, just as LTP induction apparently involves two, temporally overlapping stages, so too may LTM formation, a significant part of that process occurring during the declining phase of activation, or stage 2. If onset of stage 2 is either unduly postponed, e.g., by persisting central activation, or is too abruptly terminated, e.g., by administration of anesthetics immediately after training (chapter 3, above) consolidation may fail—in the latter case, essentially because the slope of the curve representing the subsidence phase has become too steep. (Playback was cut short.)

If LTP plays the part just described, it follows that units showing LTP may or may not also become components of a long-term (i.e., postsynaptically "fixated") memory. Some units in neocortex may do both; those in CA1, though certainly capable of LTP, may not themselves undergo the further changes supporting long-term recall. (Hence, in the macaque, the "time-limited" hippocampal memory functions reported by Zola-Morgan and Squire, 1990.) The basic function of LTP in the dentate gyrus and hippocampus may be to support the additional processes needed for the establishment of long-term memories in neocortex, in particular in those areas of "distal" association cortex (Graybiel, 1974) that are most closely related by phylogeny to the hippocampus itself.

In neocortex, the differential susceptibility of assemblies in lower layer III and in V–VI to LTP (Bindman and Murphy, 1988) may reflect the fact that they represent the two-tiered memory system described in part 1 of this chapter. A considerable part of that system, comprising sensory and parasensory areas, appears to be memory-forming on the more fluid probabilistic principle proposed by Blakemore (1974).

Beginning with parasensory or proximal association areas such as Brodmann 19, and continuing into more distal areas, in which the core elements of LTM may be laid down, the whole system is linked in essentially three ways: (1) directly in series (via layer III-to-III relays); (2) indirectly in parallel via the thalamic association nuclei (e.g., lateralis posterior and the pulvinar); and (3) recursively, over loops connecting each of the several areas with the FD and hippocampus, by way of entorhinalis and the perforant path (Fig. 4-4).

(The fronto-posterior loops that figure in "working" memory, or in the organization of overt behavior including spoken or written language, will not be considered here.)

As noted earlier, the recursive activity involved, for instance in retrieval, will consequently tend to have a reinforcing effect on assemblies in those parts of the loop in which fixation via hippocampal playback does not occur. That is, it will act, by reuse but without fixation, to strengthen memories in the loop that, *ab initio*, were established purely probabilistically.

Such playback may therefore act competitively with inputs from the sensory side. That is, it will tend to maintain the *status quo ante*—the path(s) already existing—whereas inputs from the sensory side, to the extent that they reflect changes in the organism's day to day experience, will tend to cause changes in (some of) the (same) path(s).

As I will try to show in chapter 7 here, it is the shift in balance between these two tendencies that causes some of the (nonpathological) changes in memory that often occur in old age. And because at any age, the balance is never a perfectly stable one, it means that our retrieval and recognition processes—our perspective on present events—are peculiarly sensitive to present context. (For example, when the latter is abruptly and radically changed, the result is sometimes the disorientation phenomenon called "travel memory," a transient but often upsetting disruption of normal retrieval processes.)

Hence, also, the effectiveness of deliberate retrieval or mental rehearsal in helping to make things "stick in our minds." Playback is here acting to run Blakemore's principle in reverse; i.e., to maintain past context in the face of present changes.

In motor cortex, LTM as defined here may not occur because that cortex is not reached by the projection system mediating hippocampal playback. (See Fig. 4-4 above.) The pro tem conclusion might be that motor memories, like those in sensory and parasensory cortex, are formed essentially on Blakemore's principle, without a subsequent "fixation" process.

Henry, Milner's bilaterally hippocampectomized patient, showed a normal learning curve for a mirror-writing task, but was unable to recall having learned it. He remained capable of motor or procedural learning but had lost the ability to add further to his store of "declarative" memories (Squire, 1987.)

In other words, the fundamental difference between procedural and declarative ("episodic," consciously available) knowledge may reduce to the following:

(1) That motor learning, within limits imposed by the genetically fixed final output pathways, is essentially probabilistic, and ultimately involves not the establishment of new pathways but a set of biases in favor of certain pathways already existing. And

(2) that declarative knowledge essentially involves selection of a subset of N' pathways from a repertoire of N possible pathways, where, in general, $N \gg N'$. In this sense, LTM formation amounts to new path formation. Such memories include fixated components and correspond to motivationally or affectively significant particularities in the individual's experience.

Because many of these are the result of chance, LTM is sometimes called "episodic" memory. The term is misleading (as is the term "declarative" because of its verbal connotation), since LTM also represents the type of systematized (and not necessarily verbal) recall that can result from sustained intentions or sustained feeling states. The function of hippocampal playback and "fixation" is to protect at least the key components of such memories from probabilistic erosion due to the continuing flux of experience. The fixation of those components is then what permits information that is motivationally important to us to be re-membered (*sic*).

These may be the features that distinguish long-term memory functions of posterior distal association cortex from those of all other cortical areas, including not only sensory and parasensory but also motor and much of other frontal cortex. Note, however, that in the macaque, frontal association areas 9 and 46, and mesial areas 32 and 25, do have reciprocal entorhinal connections (Fig. 4-4).

Evidence cited in the next chapter supports the conclusion that LTP can occur in many systems (e.g., the amygdala) that do not figure directly in the establishment of LTM. In other words, long-term potentiation, despite its role both in procedural and declarative learning, may not, of itself, result in LTM. It is a key but not (as many have come to believe) *the* key. As noted also, long-term memories, not being fixed *in toto*, are subject to changes over time, but indirectly, by way of changes in their unfixated components.

Motor or procedural memories are, by their nature, more dependent upon use or practice for their maintenance. However, Alkon (1989), speaking of learning in *Hermissenda*, reports that in thoroughly "trained" units, there appears to be a shrinkage in axonal volume, or some loss of collaterals.

This is a finding that deserves further investigation, particularly in higher vertebrates, since it suggests the possibility that motor learning tends to be self-limiting on the axonal or output side. If this finding were to hold true for mammals, it would explain why motor skills learned early tend to persist, and even after long neglect, can be reactivated relatively easily. For the same reason, if highly complex motor skills such as ballet dancing or piano playing are not acquired early (say in the first decade of life), they are acquired, if at all, with much difficulty later on.

It is important to note the probable role of the thalamus in the memory systems described here.

Iriki et al. (loc. cit.) were reportedly unable to obtain LTP in motor cortex by tetanic stimulation of the thalamic motor nuclei. Gabriel, Foster and Orona (1980) found that during training in an operant conditioned response in the rabbit, there was initially a focus of single-unit activity in layers V–VI of posterior cingulate cortex. However, after training past criterion, the focus of unitary activity had apparently shifted to the anterior thalamic nuclei.

The conclusion suggested by these studies is that at the cortical level, development of learned responses may primarily depend upon serial processing—perhaps necessarily so, given that the complexity of the cortex is far greater than that of the thalamus. In posterior cortex, for example, lateralis posterior and the pulvinar may act as an accessory system during the acquisition phase of learning. Once formation of a given LTM is complete, these relay nuclei may then provide a set of short-cuts, in effect prefacilitating certain of the assemblies reached via the interareal relay system, e.g., during processes such as underlie the recognition of a familiar sensory input.

The same may happen during formation of motor memories. (I am here omitting discussion of the cortico-striato-cortical loops described in chapter 6, though these clearly figure in functions of frontal association cortex, as well as in the servomechanical regulation of motor output. Apropos of this point, see DeLong (1987) concerning frontal-like behavioral functions of the caudate. The striatum may be analogous to the association nuclei, in that both can duplicate certain functions of higher-level systems, even though detailed sensory or motor learning evidently depends upon the latter. However, if this loop is partially cut, e.g., by autoimmune or other damage to the caudate, attentional deficits, compulsive behavior or tics may result, implying that it is an important link in the cortico-subcortical control system.)

Initially, formation of a motor memory may involve a succession of correction cycles, during which somatosensory input into frontal motor areas acts to refine neuromuscular performance in light of immediately precurrent results. In proportion as the resulting behavior works as intended, the amount of central activation accompanying it will tend to decline, and with it the amount or intensity of input to primary motor cortex from postcentral cortex or from the frontal supplementary motor and association areas. At this stage, elicitation of the response will occur more directly by way of thalamic (VL–VA) input.

In general then, serial processing at the cortical level may be involved in learning, parallel processing in recognition or elicitation of the already learned. This account would be consistent with (or indeed, predict) the EEG changes that ordinarily occur during training to criterion in an operant conditioned response.

The desynchronization seen in early stages of acquisition of the response reflects a rise in central activation, and corresponds to a period in which specific-nonspecific coactivation and serial processing in the cortex are resulting in formation of a sensory memory (for the conditioned stimulus or CS) and of the related motor memory (or CR). Past criterion, or when the response has reached the habitual stage, the cortical EEG response becomes highly focal, reflecting (1) a minimum of concurrent central arousal and (2) a switch in the cortex itself from mainly serial to mainly parallel processing.

It should be noted that in stressing the nonspecific or reticulocortical component in coactivation, I do not mean to discount the classical Hebbian concept involving convergence of specific inputs (on a given neuron) from two or more different sources upstream. I merely mean to suggest that in many or most cases, the critical additional requirement may be a rise, above some minimum level, in nonspecific input to the same neuron. It is this nonspecific component that is crudely imitated when tetanization is used to induce LTP.

Why Memories Arising from Hippocampal-Cortical Interaction Form a Unique Class

What emerges from this account is that so-called "declarative" or "episodic" memory—what I have here referred to as LTM—represents a cortical function that has resulted from the survival and evolutionary expansion of relations that originally existed between the *primordium hippocampi* and dorsal general pallium of amphibians.

The latter is the presumed ancestor of those distal association areas that, in higher vertebrates and man, are the parts of neocortex most closely related to the limbic system. Their primordial function may have been the retention of memories mediating the recognition of odors and thereby (if we suppose such memories to be associative), related inputs in other modalities.

It is important to note here that the amygdala and piriform cortex probably had a similar odor-recognition function, and in the rat reportedly still do (Beaulieu et al., 1988). For reasons explained in chapter 8, the corresponding memories are perhaps more narrowly specific and more directly action-related than those mediated via the hippocampal system.

Odors are highly specific but (except for intensity or approximate direction) dimensionless stimuli. There is not a precise spectrum of smells as there is of tones or colors; nor can they be sorted into sharply defined categories the way shapes can. The memories responsible for their recognition must, however, be specific, as it were to match the specificity of the molecules triggering olfactory responses in the bulb. It was perhaps to meet this special need that the hippocampus developed the "fixation" or special labelling functions described here. In higher vertebrates, it was then the extension of that labelling function part way into neocortex that gave rise to long-term memory.

With the evolution of neocortex, the prime receiving areas and immediately adjoining sectors (parasensory and proximal association cortex) came to be a system whereby finely resolved auditory, visual or somatosensory data might be recombined, in stages, to represent most commonly encountered features of the external world.

The so-called specific thalamocortical projection system was thus a misnomer, in the sense that the memories to which it gave rise were probabilistic and therefore in the nature of generalizations. As running statistical summaries of the organism's day-to-day sensory experience, they provided a basis for the generic recognition of sounds and sights. And in more distal areas such as parietal 7a and 7b, they led to a representation of spatial relations, abstracted in stages from those literally representing the position of the body in space. (An example, from the clinical literature, of unilateral "neglect" of this abstract space is given in chapter 2, above.)

The system mediating specific recognition of some of these same sensory data then arose by a kind of evolutionary improvisation, whereby projections from the hippocampus and older mesial or basal cortical structures were woven back into the newer parts of cortex. These return pathways provided a system of loops, serving to link generic components of a given memory with certain of its "fixated" or "core" components in more distal association areas.

Those components, including the mechanisms that give them their fixity, represent the parts of memory descended from the system whose chief function, in premammals, may have been a form of recall associatively linked to the recognition of specific odors. This same basic arrangement survives in the "core" components of LTM in higher vertebrates. Some of these may, indeed, be olfactory (since contrary to earlier reports, the hippocampus in man does receive olfactory input; Nauta and Feirtag, 1986); but the majority represent input in other modalities.

As I have described in chapters 2 and 3 above, these inputs are selected from a variety of data already highly generalized, and therefore of a type unlikely to be represented in the diffusely organized pallium of premammals—forms in which the specific thalamocortical projection system does not exist. The "fixation" of some of these data by way of hippocampal projections to entorhinalis and older cortex, e.g., of the lingual and fusiform gyri, is thus an evolutionary adaptation of primordial functions of the hippocampus itself.

In higher vertebrates, long-term memory then consists of lasting (mostly nonolfactory) generalized core components, associatively linked to others that are less generalized but subject to further change over time. The specificity of such memories is due to the patterns formed by these core components and the more labile referents to which they are connected.

But since the referents of a given long-term memory are susceptible to probabilistic changes in their structure or associative connections, they may alter or drop out of the pathway(s) that originally figured in its consolidation. Thus, as I have described, long-term memories can, with the passage of time, become factually incorrect, or skeletal, or virtually irretrievable. (Retrieval may still occur, though unpredictably, via the mood-related brainstem inputs outlined in chapter 7.)

The British psychologist Sutherland (1990) asks "Why do we forget?" Essentially, we do because long-term memory has this peculiar composite structure, in which highly enduring components are woven into a matrix of others that continue to be remolded by the flux of experience.

Dumont and Roberston (1986) note that certain abdominal interneurons have become part of the locust flight system, or that the lung in lungfishes has turned into the swim bladder in teleosts. The hippocampus in higher vertebrates appears to be another example of what these authors call "pre-adaptation." It represents "a characteristic that has evolved as an adaptation to one set of conditions" and "subsequently been co-opted" to adapt to others, in the ways I have tried, in this monograph, to define.

An increase both in the duration of memory and in the range of sensory information it can include has obvious adaptive advantages. As any computer user knows, it is also space intensive. And for an organism, it is as important to have some part of memory continuously updated to accommodate the shifting de facto probabilities of day-to-day experience.

The present role of the hippocampus is an evolutionary improvisation that, after a fashion, meets both of these requirements. It provides for memories of indefinite duration, but allows many of their constituent details to be represented in the continuously updated parts of the system—an arrangement that economizes on storage space, but also makes for some uncertainty in retrieval, including errors or outright failure.

To answer Sutherland—the question is not why we forget but how we manage to remember a small percentage of our experience as lastingly and accurately as we sometimes do.

Chapter 6
Applications of the Model

Part 1. The LTM Model Applied to Data from Squire. Psychological Memory Types. Motor Learning

Abstract. Explanatory power of the concepts developed here when applied to the several forms of memory described by Squire (1987). Terminological confusion in the current memory literature. "Declarative" memory (or LTM) as "uniquely accessible to consciousness" (Squire). Factors influencing the "take" time required for LTM formation. Dissociation of "procedural" from "declarative" learning, as shown in Milner's Henry.

Thompson's model of cerebellar learning. Single-neuron changes accompanying unilateral stimulation of motor neurons in fetal cord (mouse; in vitro). Temporal characteristics of postural asymmetries established in the cord, and partial blocking of these by NMDA antagonists. The basal ganglia as a regulatory feedback system vis-à-vis the cortex.

Squire (1987, p. 146, ff.) notes that "inhibitors of cerebral protein synthesis block the formation of permanent memory" (e.g., in fish, rodents, and birds, for "a variety of tasks"). Moreover, as he points out, this result can be produced "by any of several drugs having different mechanisms of action (e.g., puromycin, anisomycin, cycloheximide)."

If animals are trained during inhibition of cerebral protein synthesis, "they acquire the learning task normally but then are impaired on a later retention test." However, "if inhibition of protein synthesis is established at other times, including during the retention test itself, memory is not affected."

These results are consistent with the model of LTM proposed here. Inhibition of cerebral protein synthesis during training would not block "registration" via the hippocampal system or interfere with short-term memory. In the cortex, it would only block the postsynaptic "fixation" stage of consolidation. Similarly, if protein synthesis inhibitors were given only during retention tests or later, they would not affect an LTM already formed.

Squire cites studies indicating "that at least 10 seconds are required for synthesis of a small protein (molecular weight 10,000; 100 seconds for one of molecular weight 100,000). Some additional time would be required for proteins to move from the ribosomes . . . to the membrane of the synaptic region, where they could influence synaptic action. It would therefore appear that a protein synthesis-independent phase of memory must be capable of supporting memory for at least one minute after learning, and perhaps longer. . . . Should this protein synthesis-independent stage of memory be termed short-term memory?"

Under the present hypothesis, the answer would be yes. The protein synthesis-independent stage involves circular interaction between the hippocampus and cortex, this being a dynamic-maintenance mechanism that extends short-term memory (STM) beyond what are otherwise quite narrow limits.

Squire reports that Milner's Henry (H.M.) and rats "with hippocampal lesions" showed deficits in serial position tasks "within seconds" of presentation. He adds that Henry "can repeat a string of six digits, but never seven or eight." (However, if not distracted, he can rehearse those six digits and extend his retention of them, a form of dynamic maintenance probably involving "hold" functions of frontal association cortex.) "Rats with hippocampal lesions cannot recall the sequence of visits made to the arms of a maze only seconds earlier."

He concludes by saying that "protein synthesis proceeds too slowly to provide the basis for the putative long-term memory system that is deficient in H.M. and in rats with hippocampal lesions."

This, I believe, may be incorrect. According to the present hypothesis, the hippocampus supplies posterior cortex with two types of input that lead sequentially to LTM formation. The first is dynamic and provides the iterative support that extends STM. The second type of input is biochemical and provides the basis for LTM by way of lasting, site-specific changes in postsynaptic membrane. It is proposed that these may involve one or more immediate early genes (IEGs) and their mRNAs, resulting in release of growth factors from the presynaptic cell, and corresponding changes in receptor membrane in the postsynaptic cell.

What was lacking in Henry and in the rats with hippocampal lesions was both of these components of LTM formation—the iterative and the biochemical.

Memory Types

Figure 6-1, adapted from Squire (1987, p. 170; see also Zola-Morgan and Squire, 1990a), shows the subdivisions of memory derived from psychological studies (of normal subjects, patients with various central nervous lesions, etc.).

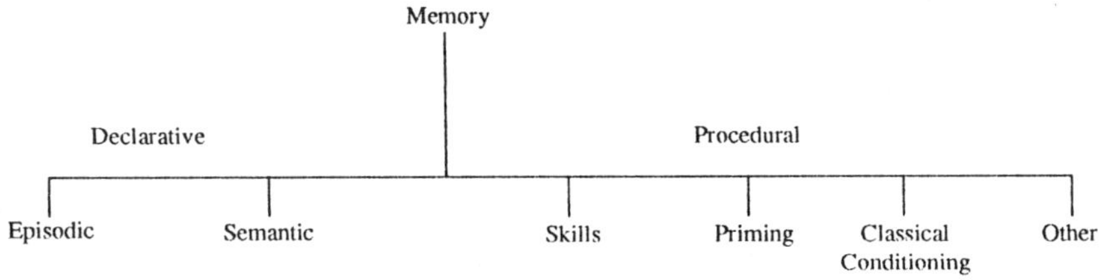

Figure 6-1. Adapted from Squire (1987, p. 170). The diagram used by Zola-Morgan and Squire differs from this one chiefly in substituting for "Other" the category Nonassociative, and under that, two items: Habituation and Sensitization. Reprinted with permission of Oxford University Press, from Squire, L (1987): *Memory and Brain.*

The subdivision of declarative into episodic and semantic memory Squire reports is due to Tulving. It is intended to reflect the fact that episodic memory derives from the subject's experience and is directly autobiographical, whereas semantic memory consists of information acquired indirectly or at second hand (and hence tends to be related to reports or words) (Squire, 1987).

This distinction is misleading, since what we commit to "semantic" memory—i.e., LTM—is as much determined by motivation as are our recollections of incidents in our own lives. The motivation is likely to be more indirect, involving items or events we "identify with," often by way of rather complex associations. Several motives may also be involved, and act additively even when mixed. We may study hard not only because we like a subject or our teacher, but because we see a career in it—and because if we don't study, we may fail and end up having to go to work.

The fact remains that the additions we make to long-term memory in this way depend on the same basic processes as those by which we remember what we were doing the day John F. Kennedy was shot. All that may differ is the "take" time, and the profound emotional aura that may have accompanied the shorter "take." This last is the kind of stress-related response that Squire says has been called "flash bulb" memory.

However, as I tried to show in part 3 of the preceding chapter, the difference between this and more slowly acquired forms of LTM can be understood in terms of tradeoffs between the time and intensity factors necessary to consolidation. If the intensity factor (affect; motivation; circulatory levels, e.g., of corticosteroids) is large enough, the time factor (or required repetitions of the input) can become negligibly small.

One exposure to the input data may consequently be sufficient to ensure their incorporation into long-term memory. And because of this intensity factor—i.e., because of the high level of reticulocortical arousal and its accompanying diffuse or sensitization-like downstream effects—such memories often include irrelevant details, as it were by a process of enforced association—which is what causes us to remember what we were doing the day JFK was assassinated.

The essential point is that distinctions between "flash bulb" and more slowly acquired forms of LTM, or between episodic and semantic memory, may have some psychological use, but as clues to the physiological organization of memory functions, they can be either useless or downright misleading.

Squire points out that the distinction between "procedural" and "declarative" forms of learning was first made in the literature of artificial intelligence, e.g., by Winograd (1975), and in cognitive psychology of the same period (e.g., by Anderson, 1976).

He notes that, as presently used, the distinction "emphasizes the idea that declarative memory is uniquely accessible to consciousness and is acquired through different brain mechanisms than procedural memory.... The two types of memory reflect different representations that cannot be substituted for one another" (Squire, loc. cit., p. 161).

Their dissociation was quite clearly shown in Milner's Henry, who could make no further additions to his "declarative" knowledge (LTM) but who showed a normal learning curve for a motor task (mirror writing; Milner, 1964), without recalling that he had this skill or how he had acquired it. In a ten-choice tactile maze, he could not memorize the turns he needed to make, but in 80 trials, showed a gradual decrease in the time he took to get through it. In other words he was still capable of "procedural" learning.

As for the other items in Fig. 6-1—priming is covered in chapter 4, part 4, above, and is misclassified in the figure. (Procedural memories relate to some form of active overt response. The memories that underlie priming relate to perception.) Classical conditioning is relatively primitive, involving the vertical (rostro-caudal) "closure" discussed in connection with affect-related retrieval in chapter 7. Learning mechanisms at the cerebellar and spinal cord levels are discussed immediately below.

Habituation (under "Other") refers to the cumulative inhibition of initially strong responses to stimuli that turn out, on repetition, to be without adaptive significance. Such inhibition is frequently of higher-level origin, as shown by the fact that habituation in the macaque can be blocked by bilateral amygdalectomy (Pribram and McGuinness, 1975). Essentially the same process underlies the extinction of conditioned responses, when presentation of the cueing stimulus ceases to have the expected consequences. The fact that extinguished responses can in general be more easily reinstated than new ones can be established suggests that extinction is not equivalent to forgetting.

Thompson's Cerebellar Learning. Spinal Cord Unitary Mechanisms and Fixation

In chapter 4, part 1, I quoted Kandel and Schwartz (1985), who defined sensitization physiologically as presynaptic, and behaviorally, as the tendency of an organism to "respond vigorously to a variety of previously neutral or indifferent stimuli after it has been exposed to a potentially threatening or noxious stimulus.... Whereas habituation is limited to the stimulated pathways, sensitization has a more widespread distribution, involving both stimulated and unstimulated pathways."

Under the present hypothesis, the postsynaptic component in memory, whether "fixated" as in LTM or dependent simply upon reinforcement through use, may become functionally less important in systems approaching the final motor paths, since these paths are of necessity genetically fixed. In pathways showing this type of convergence, a number of different "choices" made upstream can have the same outcome downstream.

Thus the specificity provided by the postsynaptic component of learning becomes less functionally important, and the presynaptic component or sensitization more prominent as we move from the receptor to the effector sides of the nervous system. Indeed (part 2, below) memories in frontal association areas, being represented in cortex relatively closer to cortical final effector pathways, may show some loss of specific structure when compared to those in posterior cortex.

It may be relations of this type that Konishi (1986) had in mind when he described presynaptic learning (in *Aplysia*) as "primitive" and "procedural." The implied distinction is between the latter and "higher" forms of learning such as LTM. Some specific differences between them, as well as some mechanisms common to both, are illustrated in the data reviewed next.

Thompson (1986) describes the conditioning of eye-blink responses (to air puffs) to a precurrent auditory tone, in the rabbit. The response is proposed to involve the circuits shown in Fig. 6-2, adapted from Thompson (loc. cit.).

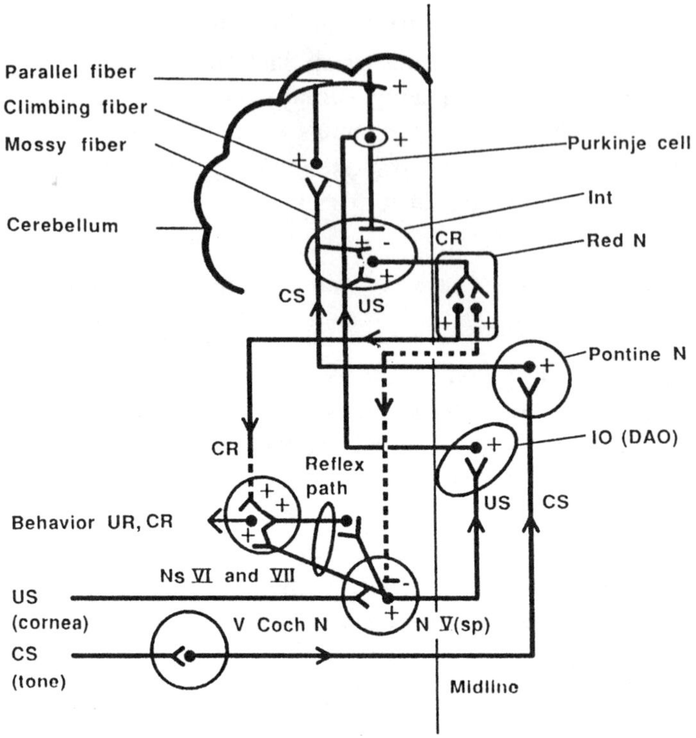

Figure 6-2. Schematic showing the circuits that may figure in the acquisition of a conditioned eye-blink response in the rabbit. The unconditioned stimulus (US), an air puff to the cornea, activates the dorsal accessory olive (DAO) by way of the spinal nucleus of V (N.V[sp]). DAO projects both to n. interpositus (Int) and to the climbing fiber system on Purkinje cells.

The conditioned stimulus (CS), a tone, passes by way of the ventral cochlear nucleus (V coch N) to the contralateral pontine nuclei (Pontine N), which project back across the midline to the mossy fiber system and, over one more synapse, to the parallel fibers in cerebellar cortex. In addition, both the CS and US are proposed to activate a circuit from n. interpositus to the red nucleus (Red N) that triggers the eye-blink response (Behavior UR, CR). The dotted pathway from ruber to N.V(sp) is intended to suggest an inhibitory feedback. Reprinted with permission of the AAAS, from Thompson, RF (1986): The Neurobiology of Learning and Memory. *Science* 233: 941–947.

The cerebellar climbing fibers, which Llinás (1987) describes as providing "one of the most powerful synaptic (junction systems) in the central nervous system," acts biphasically, to produce profound activation of Purkinje cells, followed by long-term depression (LTD). The Purkinje cell output is itself inhibitory, suggesting that the following sequence may be involved.

Initially, the conditioned stimulus (CS), by way of the ventral cochlear nucleus, contralateral pons and the mossy fiber system, sets up facilitatory foci in n. interpositus. The same input, relayed by granule cells to the parallel fibers, then causes Purkinje cells to fire, producing an inhibitory feedback whose effect is to sharpen the facilitatory foci taking shape in interpositus. The unconditioned stimulus (US), arriving subsequently by way of the spinal nucleus of V and the dorsal accessory inferior olive, may result in essentially the same sequence. That is, it will activate a second set of interpositus cells and deliver a more direct profound excitatory pulse to the Purkinje cells, causing even stronger inhibitory feedback.

The result, in interpositus, may then be to intensify the inhibitory surrounds of the excitatory foci established, in succession, by both inputs—a coactivation effect that depends on their close temporal proximity and increases their probability of becoming "associated." The latter amounts to a pro tem connection such that, with sufficient repetition of pairing, the CS alone can trigger the eye-blink response initially dependent on the US.

(Another way to view this whole process is as one in which massive inhibition acts to block out all activity except that immediately in progress. It is similar in principle to the enforced attention to some external threat that accompanies sudden intense fear.)

The foregoing, then, is a clear-and-reset mechanism similar to that provided by the rise in NE and ACh inputs to neocortex during reticulocortical arousal (chapter 4, part 3, above). If reset is followed by a relaxation phase, in which Purkinje inhibitory output is reduced (due to LTD), this may be secondary to the whole process and permissive, since the associative connection between the CS and US may already have been formed.

For this conditioning to occur, Thompson reports that the optimal interval between the CS (tone) and US (air puff) is 200–400 msec. If the interval $\leq$50 msec, conditioning does not occur. This suggests that about 50 msec are required for the CS to establish clear facilitatory foci in interpositus. If, during that interval, activation of the climbing fiber system by the US occurs, the resulting inhibitory feedback may effectively preclude that outcome, and thereby acquisition of the CR. The same applies if the CS–US interval is extended much beyond 400 msec, since the half-life of the focal activity elicited by the CS may be short.

As in other conditioning, establishment of this CS-triggered eye-blink response evidently requires some minimum number of trials. However, if the US evokes a nociceptive or startle response, it would be likely to include a reticular activating component that would augment the response, e.g., in the spinal nucleus of V or, perhaps even more so, in the dorsal accessory olive.

Nauta mentions that the central tegmental tract ("a remarkably compact bundle considering its widespread origin from the mesencephalic reticular formation") is lateral to the rostral pole of the inferior olive and is "thought to be the chief system of afferents" to it (Nauta and Feirtag, 1986, p. 190).

Hence in this system, as in the cortex, input from the RF may combine with other inputs, providing the "coactivation" conducive to associative learning. On the same principle, increasing the intensity of the US (air puff) should within limits decrease the number of trials needed to establish the CR. And if the RF is, in fact, chiefly responsible for this accelerative effect, concurrent stimulation of the central tegmental tract should mimic the effects of increasing the intensity of air puffs.

Thompson (1987) reports that "appropriate lesions" to the dorsal accessory olive result in "extinction of the behavioral CR" with continued pairing of the CS and US. This implies that, because the US can no longer supply part of the joint cerebellar input that led to an "association" between the two, the association itself dies away. The CS ceases to elicit the CR.

In other words, associative learning in this system is probabilistic and pro tem. Unlike long-term memory in Henry, it cannot become independent of the stimulus conditions that initially led to its formation. This is a critical point.

There is another. It appears that retention of the response may depend on n. interpositus, since it survives ablation of the cerebellar cortex (though the more extensive the ablation, the more prolonged is the resulting "transient loss"). As illustrated by data on the basal ganglia reviewed below, the "wiring" of nuclei tends to be more genetically predetermined and less capable of plastic change than is the case in corticated systems.

Consequently, memory formation, e.g., in interpositus, may involve not new path formation, but a relatively small number of either-or choices that can be made presynaptically. That is, conditioning can occur by way of inputs that drive one set of intranuclear synapses to the point of inducing sensitization-like changes that shift the balance of certain either-or relations in their favor, this being the basic model of motor or procedural learning proposed above.

The argument as to whether Thompson's findings represent "true" learning, or even whether they represent cerebellar learning at all (see Bloedel, 1987) really comes down to the distinctions just made. Some recent studies of spinal cord mechanisms are relevant to these.

Nelson et al. (1989) state that "the functional architecture of synaptic circuits is determined to a crucial degree by the patterns of electrical activity that occur during development. Studies with an in vitro preparation of mammalian sensory neurons projecting to the ventral spinal cord show that electrical activity induces competitive processes that regulate synaptic efficacy so as to favor activated pathways over convergent inactivated pathways."

In their study, using dorsal and ventral root ganglion cell cultures from 13-day fetal mouse spinal cord, they applied electrical stimulation over a period of 3 to 5 days, to one of two sets of spatially separated dorsal horn units having axons that converged on the same ventral horn neurons. Similar pairs of unstimulated units served as controls. The outcome was that, in the ventral horn units, the median e.p.s.p. (excitatory postsynaptic potential) evoked from the previously stimulated cell set was 10 mV, vs. 3.9 mV, for the contralateral (unstimulated) set. The median e.p.s.p. for the control cells was 5.5 mV, or roughly in between.

Taken together, these results suggest that in the stimulated-unstimulated pairs, a two-way shift occurred, in which synaptic efficacy in the stimulated cells went up and that of the unstimulated went down. The authors add that "the persistence of relatively ineffective, even subliminal synapses has been documented in the mammalian spinal cord." The implication is that if one set of ventral horn inputs is frequently used and another set is seldom used, there are corresponding hypertrophic and atrophic changes at the synapses involved. Changes in presynaptic vesicular output might account for these results. That is, we are here dealing with a form of sensitization, whose converse could result in the subliminal synapses mentioned by these authors.

Anderson and Winterson (1988) report that in the rat, repetitive electrical stimulation of a hindlimb can induce a flexion that "persists for many days." Because NMDA receptors are found in the cord, and spinal LTP has been reported, the authors treated experimental animals with an NMDA receptor antagonist (MK-801, 10 mg/kg, i.p.) 1 hour before stimulation.

Immediately after stimulation, the flexion measured in controls (n = ?) was 24.4 gms, and at 72 hrs had declined to 10.4 gms. In the experimentals, the figures were 15.6 gms poststimulation, and 10.8 gms at 72 hrs. The animals were then transected at the level of T-7 (to remove higher-level inhibition) and tested again. In the controls, flexion increased from 10.4 to 14.8 gms; in the experimentals it showed a marginal decline (to 9 gms).

These results indicate that NMDA receptors (and hence glutamate and

LTP) may figure in acquisition of the flexion response. Note, however, that in the controls, the response shows a significant decline over 72 hours (from 24.4 gms poststimulation, to 10.8 gms, or 14.8 gms following transection). This "forgetting curve" suggests that spinal learning was mediated presynaptically by LTP, or essentially by a form of short-term sensitization that died out as a function of time. That is, true fixation of the response did not occur. (The results, however, do not rule out a postsynaptic component in the response.)

Earlier work, by Chamberlain et al. (1963), showed that some consolidation time is required if experimentally induced spinal "learning" is to survive transection. The authors established postural asymmetries in rats by unilateral cerebellar lesioning. They then tested the effect on these asymmetries, produced by sectioning the cord at the level of T-7, at various postoperative time intervals. The minimal time required for "fixation" of the asymmetry was ~45 minutes. The latter might then be a measure of the time required to establish the LTP needed to support the response.

The Anderson and Winterson study indicates that spinal "fixation" declines over a period of 72 hours, suggesting that this type of learning requires repetition of the initiating inputs at some minimum interval if it is to persist. In other words, it may represent a form of sensitization, based upon LTP, whose outside duration is variably described in the literature. (Some state that it can last for months; others, such as Bekkers and Stevens, 1990, simply that it can be "long-lasting.") There is as yet no evidence that it can last indefinitely, whereas some long-term declarative memories evidently can.

With the reservations noted, the study by Nelson et al. (1988) suggests that long-term motor learning may be accompanied by—indeed may partially depend upon—losses in the efficacy of competing inputs to final output assemblies. In turn, this would help to explain why excellence in complex motor skills such as the dance or playing the piano needs to be acquired early if it is to be acquired at all.

The Basal Ganglia and Vertical Organization

Figs. 6-3 and 6-4, from DeLong (1987) and Bergman et al. (1990) show some of the loops connecting the basal ganglia with the cortex.

The striatum appears to be split (into the putamen and caudate) by fibers of the internal capsule, but the split is also a functional one. The caudate (as I recall, from a cardboard model of the brain I assembled many years ago) is relatively quite large, forming a fronto-posterior arc between the head and tail of the nucleus. Frontally (see Section No. 96, in

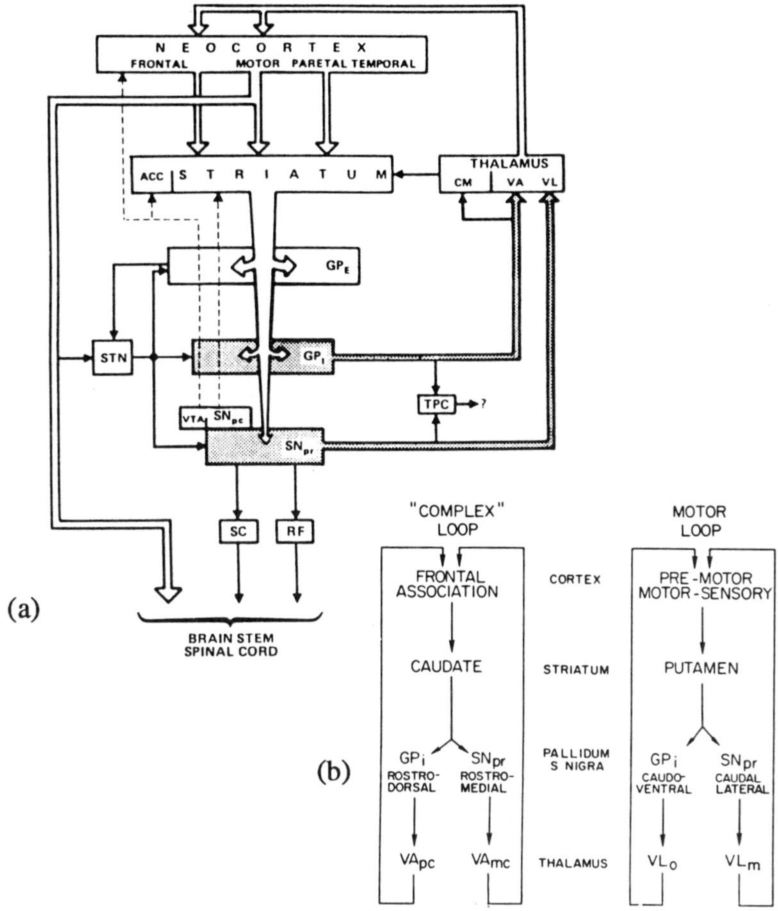

Figure 6-3. Panel (a) is a schematic showing the principal connections of the basal ganglia. Panel (b) shows DeLong's "postulated segregation of pathways from the association (complex loop) and sensorimotor areas (motor loop) through the basal ganglia and thalamus." Note that in the "complex loop" the thalamic nuclei reached by GPi and SNpr are parvocellular and magnocellular ventralis anterior, respectively, whereas the "motor loop" accesses ventralis lateralis oralis (via GPi) and ventralis lateralis medialis (via SNpr).

GPe, external pallidum; GPi, internal pallidum; SNpr, substantia nigra, pars reticulata; SNpc, substantia nigra, pars compacta; STN, subthalamic nucleus; CM, centrum medianum; VL, n. ventralis lateralis; VA, n. ventralis anterior; SC, superior colliculus; RF, reticular formation; VTA, ventral tegmental area; TPC, n. tegmenti pedunculopontinus, pars compacta. Dashed lines indicate dopamine pathways. Projections from the raphe to the striatum and substantia nigra are omitted. Reprinted with permission of Birkhäuser Boston, from DeLong MR (1987): Basal Ganglia. In Vol. 1, *Encyclopedia of Neuroscience*, ed. by G. Adelman, pp. 113–114.

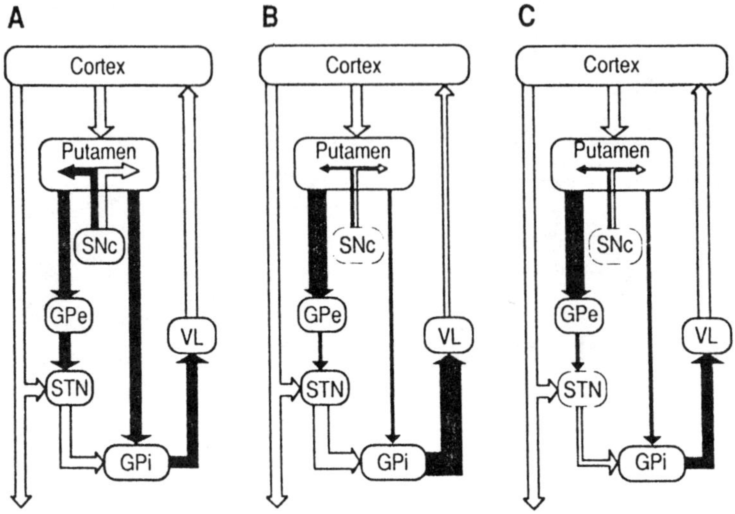

Figure 6-4. Schematic showing inhibitory (dark arrows) and excitatory (open arrows) circuits in the basal ganglia. A represents the normal balance, B the increase in inhibitory output of the putamen and GPi resulting from MPTP damage to the substantia nigra, pars compacta (SNc). In C, lesions to the subthalamic nucleus (STN) reduce the excitatory driving of that nucleus by cortical outflows, which in turn reduces the inhibitory output of GPi to ventralis lateralis, and thereby the experimentally induced Parkinsonism produced by nigral damage in B (in the macaque).

MPTP is the neurotoxin 1-Me-4-phenyl-1,2,3,6 tetrahydropyridine. The abbreviations are the same as those in Figure 6-3. Reprinted with permission of the AAAS, from Bergman H, Wichman T, and DeLong, MR (1990): Reversal of Experimental Parkinsonism by Lesions of the Subthalamic Nucleus. *Science* 249: 1436–1438.

Nauta's atlas, Nauta and Feirtag, 1986), the caudate and putamen meet in the nucleus accumbens, which forms part of the "limbic striatum" (Nauta and Feirtag, loc. cit.). It is perhaps in this area that the ventral putamen receives the substantial input from inferotemporal cortex reported by Pribram (1974).

The caudate, as might be expected from its greater extent, receives its major inputs from the association areas, whereas "the motor, premotor and somatosensory cortices project largely to the putamen. Both anatomical and physiological studies suggest a role of the putamen in more strictly motor, and of the caudate in more complex behavioral functions. In experimental animals, it has been shown that discrete bilateral lesions of specific regions of the caudate nucleus produce the same behavioral impairments seen following lesions of the areas of prefrontal cortex that project to the lesioned caudate areas" (DeLong, 1987).

The putamen has been found to be topographically organized in a way reflecting its inputs from motor, premotor and sensorimotor cortical areas. This organization is continued into the subthalamic nucleus and the internal and external segments of the globus pallidus (GPi and GPe). DeLong adds that "the segregation of influences from association cortex and sensorimotor cortex in the caudate and putamen, respectively, appears to be maintained throughout the GP, SN (substantia nigra), and thalamus" (Fig. 6-3, lower panel). The supplementary motor area, in particular, appears to be a target of striatal feedback in the circuits involving the putamen.

It should be noted that the centromedian nucleus (CM), which is part of the nonspecific thalamic projection system, forms a subloop that projects back to the striatum. The subthalamic nucleus (STN) is also part of the reticular core system, being "a continuation of the midbrain tegmentum into the forebrain" (Nauta and Feirtag, 1986).

The pars reticulata of the substantia nigra (SNpr) projects to parts of the dorsomedial nucleus (DM) and to the superior colliculus. GPi, besides projecting to the (mesencephalic reticular) pedunculopontine nucleus (Fig. 6-3, upper panel), also sends fibers to the lateral habenula. The latter is part of the ancient olfacto-habenulo-interpeduncular motor system discussed in chapter 8.

In other words, at several points—in the caudate-putamen junction forming n. accumbens, in the subthalamic nucleus, centrum medianum, and in the pedunculopontine projections of GPi—we find the basal ganglia intertwined with the reticular system. The nucleus accumbens is thus in a sense a motor counterpart of the nucleus basalis, whose role in reticulocortical activation and in memory functions is discussed in chapter 4, part 3.

Of the approximately six types of neurons found in the striatum, most are spiny, Type 1 GABAergic units, inhibitory for nigral and GP neurons. DeLong notes that the "small percentage" of striatal cells that are interneurons are "primarily the large cholinergic aspiny neurons," the system as a whole forming "an anatomical mosaic, with marked discontinuities in the distribution of neurotransmitters, enzymes and receptors" (DeLong, loc. cit.).

He adds that "it is nonetheless likely that the basal ganglia perform uniform operations upon different cortical inputs, in view of the uniformity of intrinsic synaptic organization at comparable levels of the functionally segregated pathways." It is interesting also that cerebellar inputs to the thalamus apparently follow "parallel pathways, without convergence at either thalamic or cortical levels" with the return paths from the basal ganglia.

That is, higher-level inputs from the cerebellum may represent a final motor correction system, whose signals are accordingly kept separate from those arising in the basal ganglia, which represent a more proximal stage in the higher-level generation of overt behavior.

DeLong notes that during visuo-motor step-tracking tasks "onset of neural activity in the basal ganglia... appears to lag that in the motor cortex, suggesting that these receive an efference copy of motor output."

In the earlier literature, the basal ganglia were thought to figure in postural adjustments. Later studies showed them to be particularly important in slow movements, which Parkinsonian patients have difficulty in initiating (DeLong, 1974). These features are consistent with the fact that, in this system, sustained tonic or background inhibition is maintained by the "high spontaneous discharge rate(s)" of GPi and SNpr neurons. By contrast, the tonic output of striatal units—i.e., those that mediate the inhibitory suspension of inhibition—tends to be low. "Intracellular recordings have shown them to be hyperpolarized... and extremely difficult to fire, even when excitatory postsynaptic potentials (EPSPs) of considerable magnitude are evoked by stimulation of their inputs" (DeLong, 1974).

I have gone into these features of striatal organization in some detail, because it has frequently been suggested that the basal ganglia have memory functions of some sort. The evidence reviewed suggests that, like the amygdala (chapter 8), they figure in attention. It also suggests that they exert that influence by cortical playback, an influence that, although important, is stereotyped, reflecting what DeLong calls their "uniform functions."

The latter might be defined as consisting of topographically organized inputs and outputs, each set of which is handled by wired-in algorithms of the form "not B until A" or "if C then only until G" or "if X then nothing; cancel X." The form of these algorithms presumably developed *pari passu* with the evolution of the cortical motor and sensory systems; and thanks to natural selection, they correspond to what is practicable by the motor systems finally addressed. As a measure insuring functional coherence, the same alogrithms will also, by feedback, have some power to limit the higher-level activities in which motor output originates. (Hence, for example, the concentration of return fibers from the putamen to the supplementary motor area.)

There are as yet no data indicating that these cybernetic functions include local changes equivalent to striatal learning. The fact that damage either to the caudate or to frontal association cortex can result in similar disorders does not, in short, tell us that the two systems are functionally comparable. It merely tells us that the proper functioning of one involves regulatory feedback from the other.

Part 2. Language Functions and the Nature of Frontal Memory

Abstract. Input–output relations of the frontal system, including the amygdala and DM (the thalamic dorsomedial nucleus). Its dependence upon posterior cortex for sensory information. Orderly fronto-posterior projections, such that proximal and distal association areas are connected proximal-to-proximal and distal-to-distal. Relation of these connections to the level of generality of frontal memories, and of the latter to the tactical or strategic organization of overt behavior.

The frontal system as an amplifier of basal affective and/or drive states. Effects of frontal and posterior resections compared. Circular fronto-posterior interaction during the use of spoken or written language.

The frontal lobes in primates have been as much studied as perhaps any area of the brain. For example, in the monkey, after bilateral orbitofrontal ablation, the animal

> retains all its powers of voluntary motion.... It continues to see, hear, smell and taste, and to perceive and localize tactile impressions as before.... It is also capable of exhibiting various emotions. The result therefore is almost negative.... [Yet the operated monkeys] seemed to me...to have undergone a great change.... Instead of being actively interested in their surroundings [they] ceased to exhibit any interest...sat stupidly quiet or went to sleep, varying this with ceaseless and purposeless wanderings...and generally appeared to have lost the faculty of intelligent and attentive observation.

This report, reprinted in Fulton (1951a) was written by David Ferrier in 1875. The problem it raised, as to what exactly frontal association cortex does—how it acts to organize our behavior—remains unsolved. To what extent are those organizing functions dependent on memory formation in frontal cortex?

In the frontal system there are two principal outflow channels—through area 4 to the pyramidal tract, and via orbital cortex to the hypothalamus. The former acts on the skeletal musculature. The latter mediates autonomic changes that are supportive of bodily action and strongly reflect the affective or drive states that frequently accompany it.

Being action-related, frontal functions are chiefly concerned with time, or, more precisely, timing. Hence the frontal eye fields (area 8) have derived visual functions (dependent upon postero-frontal relay), and are concerned not with the classification or other identification of visual inputs. The function of frontal cortex is to use the data available in whatever modality, as guides to the orchestration of overt behavior. To that end, units of the sulcus principalis in macaques have the visual "hold" functions described by Goldman-Rakic (1986).

Many of these functions may be wired in rather than learned, providing an automatic guidance system called into play by intended or ongoing behavior. The question then is: which parts of the frontal system are memory-forming, and in what way do frontal memories differ from those formed in posterior cortex?

Frontal Input–Output Relations

Posterior association cortex has two principal sources of input—that provided by the interareal relay fiber system, and that provided by the thalamic association nuclei. (The part played by the nonspecific nuclei is even now not well understood.) Both handle information derived from the specific thalamocortical projection system, which is to say from the sensory periphery.

The chief thalamic input to frontal association cortex is from the dorsomedial nucleus (DM; or mediodorsal nucleus). The medial magnocellular division of DM receives projections from "olfactory cortex" (i.e., pyriform, sometimes called prepyriform, cortex) and projects to orbital cortex (Nauta and Feirtag, 1986).

The central amygdaloid nucleus, which Sarnat and Netsky (1981) describe as "a continuation of the lateral olfactory nucleus (nucleus of the stria terminalis)" also projects to orbitofrontal cortex (Nauta and Feirtag, loc. cit.). The basolateral nuclei of the amygdala project to DM as well. The anterior thalamic nuclei have projections to ventromedial frontal cortex, overlapping those of DM (Fig. 7-1, below). In plate 96 of his atlas of brain sections, Nauta points out that the opercular cortex of the insula "is continuous with the rearward extension of orbitofrontal cortex" (loc. cit., p. 241) and that the latter is the only part of neocortex with direct hypothalamic projections.

These neuroanatomic features strongly suggest why it is that medial and basal cortex of the frontal system is implicated in the visceral sensations often accompanying strong emotion, in particular fear responses involving the amygdala (stimulation of which, in humans, can cause "churning" of the stomach).

Although DM also receives input from the septum and reticular formation, "no massive, circumscribed fiber bundles to [it] have ever been identified—nothing like an optic tract or a medial lemniscus" (Nauta and Feirtag, 1986, p. 130).

The functional implications of this fact are extremely important. Essentially it means that, except for olfactory inputs, frontal association cortex depends chiefly upon postero-frontal projections for its input both of immediate sensory- and related memory-data. These fiber systems, which include the classical bundles—the arcuate and uncinate fasciculi, the cingulum, and the superior and inferior occipitofrontal fasciculi (Crosby et al., 1962)—are in many cases bidirectional.

In the case of the visual system in the macaque, these are systematically arranged, as shown in Fig. 6-5, adapted from Jones (1984; some of these projections are also shown schematically in Fig. 3-2, above). Posterior visual association cortex (e.g., 19), which is proximal to the visual prime-receiving areas, projects to frontal cortex rostral to the arcuate sulcus, which is similarly proximal to premotor and motor cortex. The more distal area 21 projects to cortex of the inferior frontal convexity (the stippled area below the sulcus principalis), which is correspondingly distal to the final frontal outflow paths.

As should become apparent from the discussion below, this fact too has important functional implications, since it means that frontal and posterior memory systems are linked according to the level of generality of the data represented in each.

If we assume that that orderly progression holds for humans, and compare it with the detailed cytoarchitectural analysis by Bailey and Von Bonin of some of the same cortex—e.g., dorsal frontal area 9, and anterior and posterior cortex of the inferior temporal gyrus—there appears to be close correspondence between the two sets of data (Bailey and Von Bonin, 1951). This correspondence suggests that the operations performed by cortex in these frontal and posterior sectors may be fundamentally of the same type. That is, the layer-III-to-III relay system may have the same function in both.

In chapter 2, above, I proposed that in posterior cortex, the interareal relay system resulted in a several-stage generalization of the data relayed, thereby providing a basis for the automatic generic recognition, e.g., of forms in the visual mode, or of phonemes or other sounds, in the auditory. The referents of a given memory—the data generalized from—were then available by way of return circuits from more distal association areas to

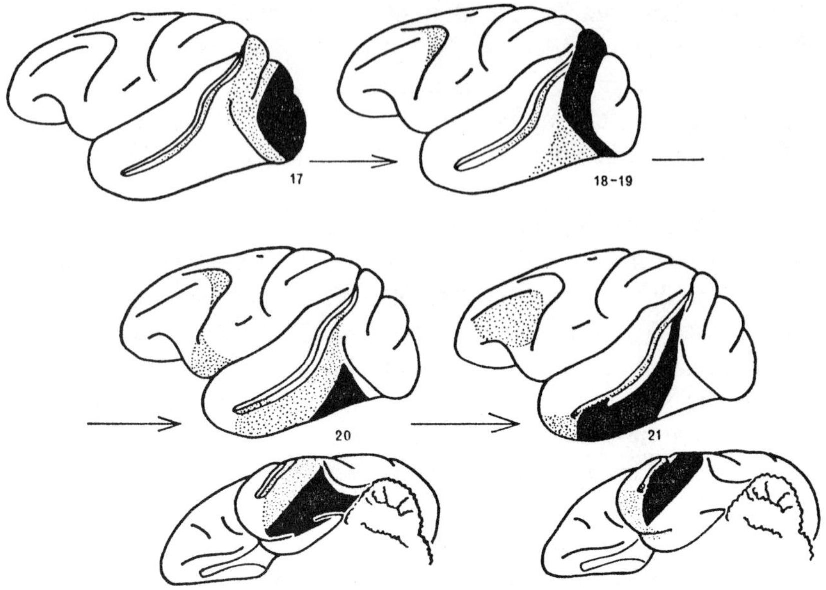

Figure 6-5. The author (Jones) is illustrating "the outward progression of corticocortical connections emanating from the striate cortex." He notes that "the progression in the temporal lobe is paralleled by a similar one in the frontal lobe." The temporal pole (area 38) projects to orbitofrontal cortex and area 20 to the amygdala. Reprinted with permission of The MIT Press, from Jones EG (1974): The Anatomy of Extrageniculostriate Visual Mechanisms. In *The Neurosciences Third Study Program*, FO Schmitt et al., eds.

those, such as visual 19, lying closer to prime receiving cortex.

Thus, whether the memory was generic and subject to further probabilistic change, or long-term and specific (because it included fixated "core" components), playback over these return circuits could retrieve some of the more particulate data that went into its formation. As suggested above, frontal cortex may work much the same way.

In frontal areas such as those shown in Fig. 6-5, the visual input-data will already have been processed and be in one of several stages of generalization, corresponding to which of the several frontal areas they reach. These sense-data may then be generalized in another way—according to their probability of occurrence in certain perceptual sequences preparatory to or directly involved in overt action.

Such generalizations are perhaps extracted from the daily flux of sensory-cum-motor activity automatically, as a concomitant of the memory formation that occurs on Blakemore's principle in posterior cortex. In other words, just as we categorize our sensory inputs in the course of normal routine behavior, we may form categories representing the sorts of things we do under the same conditions.

When the level of drive or affect rises above some hypothetical normal waking-state level, the process of memory formation frontally, (like the processes of LTM formation posteriorly) may become more stringently selective and also result in memories that are more lasting. It is in this situation that the entorhinal-frontal projections shown in Fig. 4-4 may come into play, causing the equivalent of LTM formation to occur in distal association cortex of the frontal system. In the same situation, frontal cortex and the amygdala may act jointly to tighten inhibitory controls on posterior sector activity, thereby sharpening attentional foci in the sensory systems as well. (Attentional functions of the amygdala are discussed in chapter 8, below.)

What gives the frontal system precedence under these conditions, and also determines what it selects to be at the focus of attention, is its brainstem inputs. It receives, for example, dopaminergic input from the VTA—is reportedly the only part of the cortex to do so. At least a part of that input is inhibitory (Goldman-Rakic et al., 1989), suggesting an arrangement whereby a basal drive state can set up attentional foci in frontal cortex itself. The resulting pattern then translates, e.g., by fronto-posterior relay, into corresponding attentional foci in the posterior system.

Associations in the frontal system must form in a different way from those in posterior cortex. Instead of taking shape from repeated sensory–sensory concurrences, they take shape from repeated sensory-to-motor sequences. And to the extent that that process is accompanied by powerful motivational or affective inputs, the resulting association or memory may form very rapidly.

In the frontal system, a given sensory cue or set of them does not mean (retrieve) more of the same; it means (retrieves) an action, or associated motor output. And the more closely associated the action is with a powerful drive or emotion, the tighter will the frontal inhibitory control over posterior sector sense reception be. In the extreme—in intense fear or rage—this control can so contract sensory awareness and limit posterior cortical activity generally that what we call consciousness is paralyzed or all but extinguished.

In proportion as frontal association cortex has expanded, *pari passu* with the expansion of homologous cortex posteriorly, its action-related functions have also expanded to include the strategic as well as the tactical. That is, it has increasingly come to regulate central activity that is preparatory to, rather than directly involved in, overt behavior, and the behavior itself may in some cases be indefinitely deferred. Frontal cortex also figures (as described here below) in the generation of spoken or written language, a form of behavior that, although overt, is a representation of still other forms of the overtly real and can itself be covert (as in silent speech). Consequently, though action-related by phylogeny, the frontal system figures in what are considered purely "mental" operations as well.

In chapter 2, I outlined the way in which the relay from primary somatosensory to more distal parietal cortex resulted in the generalization of literal body space into extrapersonal and finally into an "abstract" space. The latter, but not her perception of her bodily orientation or of extrapersonal space, was selectively damaged in the stroke patient described by Caramazza and Hillis (1991).

In frontal association cortex, the interareal relay of data pertaining to sequences of (literal bodily) action may result in a similar generalization-abstraction cascade, such that the temporal structure of a range of actions or events becomes probabilistically summarized in a set of "masters." These are essentially generalized flowsheets, and are extrapersonal in the sense of being blueprints for actions or events in which the subject himself may or may not participate. As such, however, they have real world referents, being experientially derived representations of real world actions or events.

When memories of this type, in turn, are processed for features they may have in common, the result is a set of third-order memories representing purely structural or analogical resemblances between events or actions that may otherwise be quite dissimilar. They thus have portmanteau functions similar to those of the parietal memories representing maximally generalized or "abstract" space.

The latter, because virtually context free, can come to serve as an all-purpose container or coordinate system in which anything, including nonspatial items, can be arranged—for instance, auditory tones (in scales, from down to up) and aims, ideals or social status (from high to low). We tend to think of such metaphors as artefacts due to our use of language, when in fact it is the other way round; we speak in these forms because they arise directly out of the way our memories are organized.

Frontal memories of the same type can serve as abstract templates for an indefinite range of operations, including verbal or mathematical ones—operations that, however remotely, represent transforms of the logic derived from our experience of actual events. But because frontal memories also have powerful motivational or affective associations, essentially as a result of the vertical organization of brain functions described in chapter 7, they tend, even more than memories in posterior cortex, to retain their tags of primitive feeling. Indeed, it is such basal feelings that drive behavior.

This relation has the odd consequence that many external events, even when they do not directly concern us, may stir up our emotions and a corresponding urge to do something—an aura of urgency that may grow in proportion as the events continue and we remain unable to change them. It is by a mechanism of this type that we come to "identify" with peoples or causes that may in fact have nothing to do with us. For the same reason, social awareness, sensitivity to the feelings of others, is believed to be a frontal function, most usually (and perhaps significantly) of the right or nondominant hemisphere. It is typical of frontally lobotomized patients that they frequently lose the capacity for this kind of involvement. They don't "identify," or at least not for long.

The frontal system, one may say, is not a "seat" of emotion or motivation, but a system in which their effects are amplified. Much of that amplification may be due to the amygdala, making its relation to the frontal system analogous to that of the hippocampus to posterior cortex.

As an amplifier of basal states, what frontal cortex produces is anticipatory "sets"—an aura of drive or affective intention that (because related to or calling for action) concerns the future. Thus while the cortical representation of pain is reported to involve the somatosensory area and the cingulate gyrus (Talbot et al., 1991), frontal cortex is apparently what amplifies continuing pain and gives it its ominousness.

In patients with intractable pain, it is as if the incessant urge to do something about which nothing can be done deepens into a horrible certainty that more and worse is to come. Such patients are often relieved by frontal lobotomy. They still report feeling pain and still know what it prophesies; they are simply no longer as oppressed by the sense of futurity that goes with it.

Memory formation in frontal cortex might be inferred, then, to depend upon a three-way associational process, relating certain sense-data relayed from posterior cortex with certain event sequences or action templates (some innate, some learned), and both with certain concurrent basal states.

A frontal memory is consequently a different sort of package than one in posterior cortex. It consists, essentially, of a brainstem-connected, action-related algorithm, in which the amount or detail of sense-data called upon need only be the barest minimum that will provide unambiguous links in a potential behavioral sequence. The sense-data relayed to frontal from posterior cortex may thus be roughly equivalent to what ethologists call sign-stimuli—triggers to action, steering cues.

Frontal memories resemble the instructions in a computer program, consisting of timelike command sequences that draw much of their energy from elsewhere, call upon data represented elsewhere and initiate actions finally carried out elsewhere. The system seems to be hierarchically organized, such that higher-level control commands can, for example, cancel any given if-then sequence, and substitute another or others. (Hence the importance of frontal cortex in divided-attention tests. It enables us, as electronikers say, to "chop.") Except during mental rehearsal, these are real-time sequences, contingent on what actually happens.

The features distinguishing frontal from posterior cortical memory reflect the fact that the number of things we may know about a given situation is likely to be far larger than the number of things we can directly or even indirectly *do* about it. At the same time, depth of knowledge may be important in that, by postero-frontal playback, it may critically influence the executive orders finally "decided on" or issued frontally.

The greatest expansion of representational memory has consequently occurred on the sensory side, or in posterior cortex. But just as posterior cortex sorts its inputs into categories of several orders of generality, frontal cortex, acting by way of intracortical relays of the same type, sorts its outputs into categories also of several orders of generality.

(An output is a procedure; similar procedures can be generalized from, producing rules applicable to the set; experience with procedures of different sets can then lead to rules as to which go with which and in what order. The rules of syntax, discussed below, are apparently built up in this way.)

It is this essential similarity in processing that we may see reflected in the orderly pattern of postero-frontal relays shown in Fig. 6-5 from Jones (1974). It is an arrangement that would make for functional consistency, in that it would connect visual data of a given order of generality with frontal "rules" of the same order of generality.

Because, to become operative, memories in frontal association cortex depend upon some minimum of input of sense-data from posterior cortex and also upon some minimum of directive, dynamogenic input from the limbic system and brainstem, they are, in themselves, almost contentless—like a computer program with no one at the keyboard and the power turned off. So while earlier experimenters were able to elicit "sparks" from stimulating area 18, or visual images from 19, in human subjects, they were puzzled to find that many areas of frontal association cortex were "silent" to electrical stimulation (Fulton, 1951; 1951a).

These relations are perhaps clearest in the case of syntax, which consists of structural templates representing real-time verbal sequences.

They are all-purpose in the sense that any given grammatical form can be filled by *N* nonidentical sets of words. A given syntactical form can also appear directly, or under transform, in a variety of larger sentence structures.

Here, the hierarchical organization of frontal memories may figure, since this higher-order organization provides a set of master schemata into which substructures can be fitted—an arrangement consistent with the fact that sentences having the normal range of syntactical variety reportedly cannot be generated by simple Markov processes.

It is just here that the difference between frontal and posterior memory functions most dramatically stands out. For while the rules of syntax number in the tens, our "expressive vocabulary" involves verbal memories numbering in the thousands, with direct or connotative referents numbering thousands more. The same is true of our actions generally. Essentially the same plan of action can, with minor modifications, be tailored to fit a number of quite different situations, each involving different sets of sensory inputs as guide data.

If the account of frontal memory functions is correct to this point, it would predict that the posterior cortical layer-III-to-III relay system (Murray and Mishkin,1985) is organized and operates in much the same way in frontal association cortex, and results in a similarly two-tiered memory system (chapter 5, part 1).

In contrast to visual neurons, which develop preferential responses to inputs they have seen most often, frontal cortical neurons may develop preferential responses corresponding to behavioral outputs that have worked most often. Here, reinforcement will depend upon correction cycles, or iterative feedback from many levels including posterior association cortex and the striatum. A further correction cycle may involve somatosensory input from area 3 to layer III in motor cortex, as described by Iriki et al. (1989).

In much of frontal cortex (as in much of posterior), LTM may not be necessary; unfixated probabilistic learning may suffice. As I have tried to show, memories in posterior cortex represent a several-stage generalization of structural features of the input, in some cases including temporal structure. The order in which we commit these data to memory or later recall them is not important (making them often hard to place in the longer time stream of our lives).

The spatial features of posterior memories frequently are important; hence the representation of space itself in the parietal lobe and the existence of "place" neurons in the hippocampus. The latter, as would be predicted under this hypothesis, are reportedly more generalized or less "sensory-bound" than those in found entorhinalis, and are also activated at longer latency (Quirk and Muller, 1989).

However, audition is more often sequential than punctate; visual cortex contains motion detection and relative motion detection units; tactile sequences (vibrations; taps) or vestibular inputs (Guldin et al., 1989) activate parietal units. So we would have to say that the contents of posterior cortical memory, although heavily biased toward the retention of structural features of sense-data, are not exclusively so.

In the frontal system, this order of emphasis is apparently reversed. Frontal memories, though having primarily to do with structure in time, must of necessity have spatial referents, many or most of which may be accessed by the reciprocal transcortical pathways discussed earlier. Tasks, for instance, whose performance depends upon complex visuospatial data may require frontal access to inferotemporal cortex, as shown in the studies reviewed by Squire (1987, pp. 226–228.)

Conclusion: Concerning Language

The inputs to frontal cortex from the amygdala and DM, and from dopaminergic projections of the VTA, as well as its hypothalamic and visceral connections via the orbitofrontal area, are clearly consistent with its role as an instrument for the construction of affect- or drive-related behavioral sequences.

In the macaque, prefrontal cortex shows some regional specialization. For instance, the inferior frontal convexity (ventral to the sulcus principalis, or approximately area 45) seems particularly to be involved in object alternation or reversal tests, and go-no-go discrimination. Lesions to this cortex lead to perseverative errors.

Dorsally and more posteriorly, cortex adjacent to the arcuate sulcus appears to mediate behaviorally related associations between visual or acoustic stimuli (Squire, 1987). Evidence of this sort is not yet as clearcut for humans, perhaps because there is more latitude in the final localization of frontal associative functions of this kind.

Pribram reports that there is considerable variation in the frontal representation of language (i.e., in the right-handed) and adds that frontal cortex shows evidence for memory of very frequently used function words such as "the" (Pribram, 1989). Since "the" has next to no meaning and innumerable structural uses, it almost qualifies as a template memory, like a conventionalized bolt in a blueprint. Patients with Broca's aphasia tend to omit function words.

Besides being forced to use telegraphic speech, some Broca's aphasics reportedly have defects in "receptive grammar" as well. Such patients have been found to have difficulty in understanding sentences that are grammatically complex (Nauta and Feirtag, 1986, pp. 88–89; the finding referred to was made by Zurif and his colleagues). That is, in some Broca's aphasics, the system that keeps track of the temporal shape of sentences has been so damaged that iterative processing over the fronto-posterior loop barely works. So unless the temporal shape of sentences is kept simple, it is lost, and meaning with it.

The essential point is that the generation of spoken or written language may depend upon the same basic intracortical mechanism(s) as does the rest of our overt behavior, and therefore provides insights into the nature of frontal memory. (Even the fact that language is symbolic is immaterial here, since a good deal of other human or animal behavior is also symbolic.)

In a recent paper (Fair, 1989), I put that idea to quantitative use. The paper presented a mathematical model that computed the rate at which we "use up" our vocabulary (i.e., in extended discourse, over a range of topics). It started with a concept derived from a study by Ojemann and Mateer (1978) that showed the existence of syntactical and short-term memory registers in the perisylvian region of waking human subjects. The concept was "rollover vocabulary"—a clear-and-reset mechanism that periodically replaced the contents of our short-term verbal registers with a new "chunk" (more or less the way we "chunk" our other behavior).

The model assumed that, like many other actions (including planned or merely imaginary ones) language involves the supplementary motor area. Geschwind (1974) reported that to be the case; and in their cerebral bloodflow study, Lassen et al. (1978) showed that this area was activated when the subject spoke, or read either silently or aloud, or when he was asked to imagine, without performing, a finger movement.

The model further assumed that speech or writing depend, like our overt actions generally, on circular interaction over fronto-posterior loops, connecting frontal memories (e.g., of Broca's area, related to syntax) with posterior memories (e.g., of Wernicke's area, related to words). In this way, for instance via the arcuate fasciculus, an iterative best fit was obtained between the words one had available for use on a given topic, and the syntactical forms one had to fit them into.

Using this model, I was able to develop an expression representing the rate at which the average writer deployed his expressive vocabulary, and the asymptote approached as that vocabulary "ran out."

The expression gave a good fit with the rate of appearance of previously unused words, both in extended samples from a single author, or in pooled 2000-word samples, drawn from 500 writers working in 15 different genres. And when run out on the computer to a theoretical sample size of 2×10^9 running words, it gave an asymptotic value for the total vocabulary in English that was within 4.33 to 0.71% of the word count for the then-current Oxford English Dictionary.

The object of this exercise was to show that, by treating language functions as differing in no essential way from the rest of our overt behavior, one could convert that model into a set of equations giving accurate linguistic results.

From a physiological standpoint, the key question is whether the reciprocal connections between entorhinalis and frontal association areas shown in Fig. 4-4 mediate LTM formation in those areas or their homologues in man. Neither the angular gyrus nor the frontal speech areas seem to have equivalents in nonhuman primates. However, frontal area 46 in the monkey (which has entorhinal connections) is comparatively close enough to the human speech area (44–45) to make one wonder if the latter may not have long-term memory functions; and whether these may not be a major reason why we communicate in sentences rather than (or as well as) by grunts and screams.

The more basic point is that the frontal system may have generalizing functions similar to those of posterior association cortex. Moreover, the *level* of generalization in each system is related to that in the other, as shown by the systematic arrangement of fronto-posterior connections illustrated in Fig. 6-5. There is even a systematic relation between the cytoarchitecture of the cortical areas connected in this way—that of distal association cortex, both frontally and posteriorly, being of the "generalized eulaminate" type.

So just as the parietal representation of space becomes generalized (and can also become "abstract" in the sense defined and illustrated in chapter 2), the frontal representation of time—the coordinates determining action sequences—can become similarly generalized. And instead of the master categories established posteriorly ("dog" = subcategory of four-legged animals = subcategory of living things, etc.), what one finds frontally is master categories related to sequencing. Via this hierarchically organized system, the essentials or over-all form of what one wants to say or do, are addressed first, and the subsequences are then called that best fit the form—each stage in this cascade involving fronto-posterior interaction.

It was these relations that I was trying to illustrate with my model of sentence generation. Note that time, in language functions—when we write or speak or talk silently to ourselves—is now partly abstracted from real time, or the timing we need, for instance in playing a racket game. It is time, abstracted to mean sequential order, without regard to the real-time intervals between the elements ordered. It is thus analogous to the "first-half, second-half" abstract space of the lady described in chapter 2.

Chapter 7

Affect-mediated Retrieval and the Vertical Organization of Memory Functions. Vestibular Disorders. Memory in Old Age

Abstract. Korsakov's Syndrome: its relation to the problem of affect-driven retrieval. Relation of this syndrome, and of vestibular disorders, to the vertical organization of memory functions. An unsolved problem: how are memories time-coded? Neuronal losses and the changes in memory structure that occur in old age.

In this chapter, I will take up the question of affect-related retrieval, considered in the light of clinical data relating to Korsakov's syndrome or to the cognitive disorders that can result from brainstem damage affecting the vestibular system. Both reflect a form of vertical central nervous organization that, in the case of long-term memories may serve to link them to their basal (motivational or affective) states of origin. The chapter closes with a discussion of the mechanisms that may lead to the nonpathological changes in memory that tend to occur as we age.

Affect-related Retrieval

Retrieval presents a complex problem because of the several types of association that seem able to elicit it. So far, I have not considered the question of how retrieval by emotional or motivational association may occur, or in what way(s) that process may differ from retrieval via the more purely factual associative connections that in posterior cortex link the items retrieved.

We have seen how the hippocampus may act as a transducer by which the sensory events occurring in a given motivational or affective climate can become incorporated into LTM. What is not at all clear is how this transduction might work in reverse. Why is it that a given long-term memory reactivated in the present can also "recall" the climate of feeling or intention in which it was originally formed?

We know that this does happen; that in general, intentions generate the appropriate related emotions, or vice versa, and that both can result in related long-term memories. We also know that affect-related retrieval can easily occur, as in reminiscent interludes, in which memories are "brought to mind" not by a current activity but by a current mood—as in relaxed states, when we allow our thoughts to drift in whatever direction our feelings may take them.

The implication is that cortical long-term memories must be reciprocally connected in some systematic way with parts of the brainstem such as the ventral tegmental area (VTA) in which many of the corresponding drives or feeling-states arise.

The extensive literature on self-stimulation suggests that in fact the latter are most powerfully represented at the brainstem level, and can be tapped in major conduits such as the medial forebrain bundle. It is interesting, in this connection, that the VTA has only recently been identified in man. It appears to coincide with Gudden's ventral tegmental nucleus (Paxinos et al., 1989).

There must, in other words, be a set of rostro-caudally running loops, such that any cortical long-term memory has its brainstem "tags" or conversely—that the tags have corresponding higher-level memories. It is important to note that this must be a many-to-few relation, since the number of our basic drives and related feeling-states is comparatively small, while the number of lasting memories we are capable of forming is indefinitely large.

It follows then, that a great many long-term cortical memories may have similar or identical brainstem tags, representing a common motivational/affective origin. Conversely, brainstem structures mediating a given intentional or affective state may be able to access a relatively large number of memories, some of which may have no other relation to each other than this one. The tagging that occurs in one direction (caudo-rostrally) is divergent, in the other convergent.

This arrangement has several familiar behavioral-psychological consequences. One is that, since thinking usually involves our calling upon memory, some of the memories called upon may carry emotional tags that are (a) unrelated to the matter in hand and (b) so powerful that they set up competing cortical activity whose direction is determined by emotional rather than factual associations.

Mood-driven reminiscences are another form of the same phenomenon, occurring simply by default, e.g., at the end of a day, when we are tired and relaxed and thinking about nothing in particular. Like fantasy or rationalization, they represent a state in which the system is being run from the bottom up. (In this connection, it is interesting that Beracochea and Jaffard (1988) report that electrolytic lesions to the mammillary bodies in mice impair what they call a "primarily automatic form of memory.")

Korsakov's Syndrome

By what pathways does that kind of retrieval occur? Clinical data relating to Korsakov's syndrome suggest some of the cortico-subcortical circuits that may be involved. This syndrome is thought to result primarily from alcohol-induced damage to the mammillary bodies, though many cases have been found, at post mortem, to have had lesions in the dorsomedial thalamic nucleus (DM) or, more rarely, in the anterior columns of the fornix. Thalamic lesions may include the medial pulvinar and laterodorsal nuclei; and in some cases hypothalamic nuclei other than the mammillary bodies may be affected (Signoret, 1985).

It is not the hippocampus (as was formerly thought) but the subiculum that projects, via the post-commissural fornix, to the mammillary bodies (Swanson, 1987) as well as to n. accumbens and the anterior thalamic nuclei (Nauta and Feirtag, 1986). The mammillary bodies receive input, via the mammillary peduncle, from the VTA and "from visceral sensory and gustatory centers of the medulla" (Sarnat and Netsky, 1981). They also receive "massive inputs from large parts of the septal region and hypothalamus" (Swanson, 1987). The septal input chiefly arises in the medial septal nuclei and diagonal band area, and is carried by fibers of the precommissural fornix.

Thus the hippocampus has two indirect links to the mammillary bodies; from the subiculum, which is transitional from entorhinalis and (like CA1) receives Schaffer collaterals, and by way of its projections to the lateral septal nuclei. (The latter, according to Swanson (1987) is the "one major subcortical output of the trisynaptic path.") The septum is known, from self-stimulation studies, to be a powerful "reward" area and also has two-way connections with the VTA via the medial forebrain bundle.

The mammillotegmental tract, described by Sarnat and Netsky (loc. cit.) as "lightly myelinated," projects to the dorsal and ventral tegmental nuclei in the midbrain (i.e., the VTA; see Paxinos et al., 1989); to the interpeduncular nucleus, and to the pontine gray matter. (The interpeduncular nucleus is phylogenetically old, was found by Thompson (1960) to figure in learned avoidance responses in the rat, and might be classified as "limbic motor.")

The mammillothalamic tract is larger, "heavily myelinated" (Sarnat and Netsky, loc. cit.), and projects to the anterior thalamic nuclei. These nuclei, besides sending projections that overlap with those of DM to ventromedial frontal cortex (Squire, 1987), project to the posterior cingulate

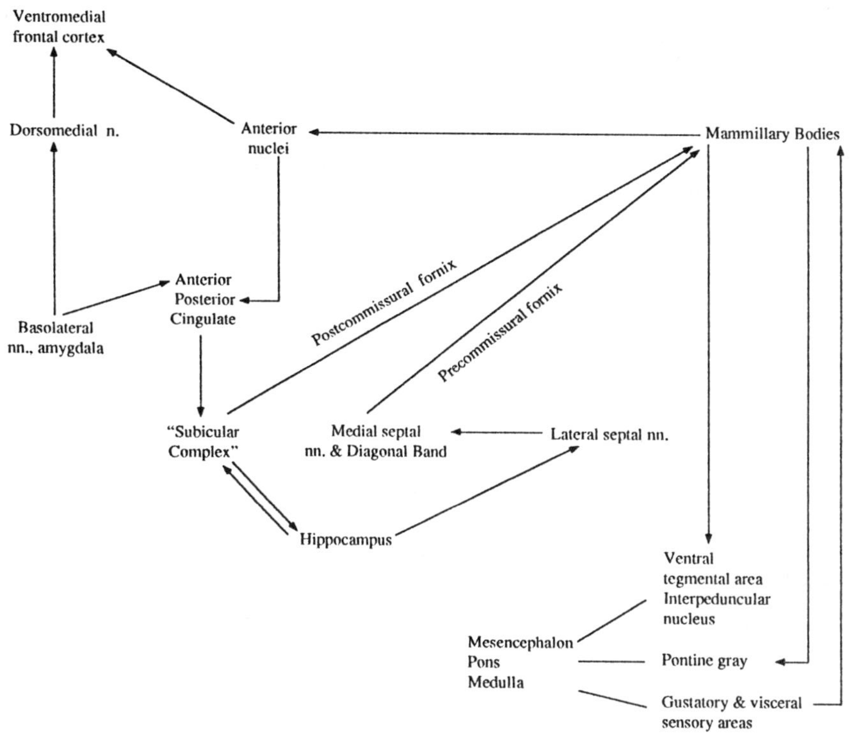

Figure 7-1. Schematic showing some of the principal connections of the mammillary bodies. See text for discussion.

(Vogt et al. 1979). Finally, the cingulate projects to the striatum, the thalamus, to several neocortical areas; and back to the hippocampus by way of the "subicular complex," as do the anterior nuclei (Swanson, 1987; the "complex" consists of the classical divisions, the pre- and parasubiculum and subiculum proper. It lies in the parahippocampal gyrus, shown in Fig. 3-1).

The hippocampus is thus incorporated into at least three major loops. The first one, discussed at length in the text, connects it reciprocally with association cortex via the subiculum and entorhinalis. The second connects it, by way of the lateral and medial septal nuclei, the mammillary bodies and the anterior nuclei, with the cingulate.

A third loop, intersecting this one, runs rostro-caudally from the mammillary bodies to the brainstem, with afferents to the midbrain tegmentum and pontine gray, and efferents, via the mammillary peduncle, from the medulla. Some of these connections are shown in Fig. 7-1.

According to one report, the mammillary bodies "are sharply circumscribed from the rest of the hypothalamus" (Nauta and Feirtag, 1986) and have little connection with other hypothalamic nuclei (contrary to the citation above, from Swanson, 1987). The mammillary bodies appear to be nodal components of a distinct, partially segregated system, in which they serve as an interface between major structures at the limbic level and related structures at the medullary, pontine and mesencephalic levels.

The motivational or affective weightings given to concurrent sensory inputs may depend critically upon these circuits; and, via the subiculum and hippocampus, it is such weightings that may determine what is or is not to become incorporated into LTM. The implication is that these weightings must also be coded into memory at some point(s) in these circuits at the brainstem level, if retrieval of a given long-term memory at the cortical level is also to be accompanied by reactivation of some of its motivational or affective states-of-origin; or conversely, if a given basal state is to evoke related memories in the cortex.

The further implication, from clinical data on severe Korsakov cases, is that when these circuits are cut in the mammillary bodies, coded data, representing the motivational or affective correlates of much of LTM, become partially or largely unavailable to the hippocampal-neocortical system. In proportion, the capacity to add to long-term memory will be diminished.

It is reported that Korsakov's syndrome produces an amnesia that is apt to be particularly acute for remote memories (Squire, 1987, p. 217; for the contrary, and more usual view, see Weintraub and Mesulam, 1985). A study by Shimamura and Squire (1986) of patients with Korsakov's syndrome, was summarized by Squire as follows:

> No correlation appeared between between the overall severity of anterograde amnesia (as measured by nine tests of new learning ability) and the severity of remote memory impairment (the 1940s to the 1970s). However for the most recent decade covered by the tests (the 1970s), the severity of anterograde amnesia did correlate significantly (r = +.36) with the severity of remote memory impairment.
>
> Test questions concerning the most recent decade presumably measured the anterograde amnesia already present, or developing, during that time period. Prior to the 1970s no correlation was found between remote memory performance and new learning ability (1960s: = .16; 1950s: +.12; 1940s: −.09). *Taken together, the results suggest that the ability to recall remote events depends on mechanisms distinct from those required for new learning and*

for the recall of recent events... (italics added).

I would interpret these results as follows:

Remote memory in these patients was impaired, due to the combined effects of the (naturally occurring) changes in memory described in chapter 5, part 3, and of mammillary damage, which acted to block affectively prompted retrieval. The anterograde amnesia, developing in the 1970s resulted primarily from impairment of *registration*—that is, from want of weighted inputs from the mammillary bodies to the hippocampal system, such as would normally tag certain new data as motivationally or affectively significant, thereby making them probable candidates for incorporation into LTM.

The severely afflicted Korsakov patient, being shut off from the sources of affect-mediated recall, tends to lose touch with his own early history as well as with his present feelings. He loses touch, in short, with who he *is* and also has trouble remembering the temporal order of recent events. In proportion, his cognitive processes slip out of gear with reality and he tends to confabulate.

Problems with time sequences, which one would expect to be frontally related, were reported in the older literature to occur in some patients with lesions to DM; and Signoret (1985) cites a report in which 11 out of 14 Korsakov cases also had thalamic damage. These eleven patients confabulated, the other three did not, suggesting a clear connection between confabulation and release from frontal inhibitory control (cf. Rubin, 1989).

A study by Janowsky et al., 1989, showed that on the Wisconsin Card Sorting Test, Korsakov patients ($n = 7$) "achieved significantly fewer categories than the 5 other patients with amnesia." They also resembled (7) frontal-lobe lesioned patients in impaired performance on the Initiation–Perseveration subscale of the Dementia Rating Scale.

The authors note that these frontal lobe symptoms find support from "recent radiological (CT) findings of frontal atrophy" in Korsakov patients. Apart from the probable involvement of DM (in the cases studied by Janowsky et al.) it should also be noted (Fig. 7-1) that some frontal input from the brainstem, via the anterior nuclei, would also be impaired in this disorder.

The loss of sense of self in severe Korsakov cases is sensitively described by Oliver Sacks in his popular book *The Man Who Mistook His Wife for a Hat.* His patient "William," besides no longer really knowing who he was, had become unable to "identify" with the people and the happenings around him. It was this double isolation (Sacks felt) that drove him into incessant confabulation, making him continually reinvent a self to do for the one he could no longer find, and contrive elaborate scenarios to explain events he could no longer properly keep track of, or plans he couldn't hope to carry out.

Sacks describes him as "blathering along... in his excited, dead way, insensitive to truth, to reality, to propriety, to everything—insensitive too, to the manifest distress of the living brother before him. It was this which convinced me... that there was some ultimate and total loss of inner reality, of feeling and meaning, of soul, in William" (Sacks, 1987).

Damage to parts of these limbic circuits other than the mammillary bodies—for example, bilateral fornix section or bilateral cingulotomy—can produce similar but seldom as severe memory disorders (Signoret, 1985; Squire, 1987). It seems to be the position of the mammillary bodies, as an interface between the two sets of loops described above, that makes them so critical. For once the system is sufficiently damaged at that interface, it fails; and no other existing systems or pathways seem able to compensate for the resulting functional loss. Certain Jacksonian levels seem to have been permanently disconnected.

One could say that the difference between Sacks's William and Milner's Henry was a difference in the depth at which the roots of long-term memory, and so of self-awareness, had been cut. Henry, despite the pain his memory disability caused him, still knew who he was and did not confabulate.

Since in Henry, and also in Squire's R.B. (Squire, 1987), remote memory was preserved, it is clear that retrieval of long-term declarative memories does not depend on an intact hippocampus; only the establishment of new ones does. However, if such memories can be retrieved not only as a result of intracortical activity, but also as a result of subcortico-cortical inputs representing their emotional associations or "tags," how is that tagging accomplished?

May it depend, for example, upon parallel representation of the "tagged" data (with more tag than data) in limbic structures such as the cingulate? But the cingulate, and many of its connections with the cortex and the hippocampus, presumably survived in Sacks's patient William. So tagging, at this level, is unlikely to be the answer.

The key, then, may lie in the loop connecting parts of the brainstem with the mammillary bodies. So long as those circuits are undamaged (as was the case in Henry), the vertical integration of data concurrently being processed at several neuraxial levels can evidently still take place.

Lacking a hippocampus, he could not add to his store of long-term memories because circular interaction over the hippocampal-neocortical loop was no longer possible. But the other routes shown in Fig. 7-1 remained—in particular the rostro-caudal loop linking several levels of the brainstem with the anterior nuclei and cingulate by way of the mammillary bodies. (Sarnat and Netsky, 1981, p. 343, describe the cingulate as having reciprocal connections with many cortical areas, making it "therefore the principal intermediary between the hippocampus and neocortex.") It was perhaps these circuits that enabled Henry to retrieve preoperatively established long-term memories by way of their emotional associations (as well as intracortically, by way of their experiential or logical associations).

The fact that declarative memories may have these reciprocal brainstem connections does not amount to saying that all apparently rational recall is therefore irrationally motivated. It merely means that there tends to be a systematic relation between what we think and feel. It is a relation that arises automatically in a system in which any higher level event has its lower level concomitants (or tags). In physiological terms, it is a form of organization that acts to maintain functional coherence between levels that are functionally quite different.

Gastaut seems to have had a similar principle in mind when he proposed long ago (1958) that, during conditioning, memory-formation (he called it "closure") might occur at several neuraxial levels, from the brainstem to the neocortex. Under this view, a given long-term memory would consist of a stack of interrelated elements, in which the simplest and most basal would be represented caudalmost. (As mentioned below, the simplest form of "closure" would then be the type that results in classical Pavlovian conditioning.)

Because of the topographic order conserved in the "stack," these caudalmost elements, representing the rather small class of possible motivational and affective states and their several combinations, would then automatically be inclusive of higher-level memories representing the much more diverse classes of experience in which such basal states may originate.

The whole system might be visualized as an inverted cone, in which successive sections parallel to the base of the cone represent successive levels of central nervous organization. The area of a section lying close the vertex of the cone (here the bottom of the system, or brainstem), would represent the relatively small class of basal states and their relatively few combinations, of which the organism is capable.

The area of the base of the cone (here the top of the system) would then represent the potentially much larger number of cortical memories that might be included in one or more of these motivational-affective classes by vertical "closure"—that is, by way of experientially established two-way links, connecting sets of points in the planes of the several sections from top to bottom. Links of this type—e.g., connecting the sound of a bell with the imminent appearance of food—are perhaps what underlie classical conditioning (Fig. 6-1 above).

Such an arrangement, if in fact it is the actual one, would make for a durable kind of inner consistency. Our conscious recollections would tend of necessity to retain their original "tags" of feeling or intention, so that, happily or not, we would go through life with an abiding, deep sense of personal continuity.

It is that sense that some Korsakov patients appear to have lost, essentially perhaps because of the disconnection between the cortical and brainstem levels of this vertical system, resulting from damage to the mammillary bodies.

(I should add that not all Korsakov patients, including another case described by Sacks, fit that description. However, given the various lesions that have been reported in this disorder—especially those involving the dorsomedial nucleus and frontal cortex—that is not surprising. See Squire, 1987, pp. 181–184.)

Vestibular Disorders

The vestibular system provides another example of vertical organization. Damage to the vestibular nuclei in man can result in higher-level disorders that have no apparent relation to position-adjustment functions of the VIIIth nerve system.

Erickson (1989) lists the following:

(1) "A decreased ability to track two processes at once."

(2) A decreased ability to handle conflicting emotions—to resist falling into "confusion."

(3) "Decreased mental stamina.... For the vestibular patient, an hour or two of concentration is a special blessing, and most days 15 minutes of intellectual concentration is very fatiguing."

(4) Impaired retrieval, or "the ability to pull out information from... long-term memory... reliably."

(5) "A decreased sense of internal certainty.... They lack that 'ah-hah, I've got it now. I see the big picture.' "

(6) "A decreased ability to grasp the large whole concept... to see the forest for the trees."

In macaques, the vestibular nuclei reportedly have direct projections to VPL oralis, which projects to area 4. In that cortex, 39 out of 158 units tested responded to vestibular stimulation, 8 of them at latencies of <8 msec (White and Brinkman, 1988). In the squirrel monkey, using retrograde tracer techniques, Guldin et al. (1989) obtained evidence that "vestibular input to the thalamus is widespread over several nuclei: the ventral posterior inferior nucleus, the medial part of the ventral posterior nucleus, as well as part of the oral and medial pulvinar."

The "cortico-cortical interconnections most outstanding were the dense efferents and afferents to area 3a." Parts of parietal area 7 (to which 3a projects) contain "a small amount" of units responding to semicircular canal stimulation. What they describe as "parieto-insular vestibular cortex" receives input from the frontal eye-fields and cingulate cortex.

Comparing these neuroanatomic data with the list of symptoms presented by Erickson (1989), it is clear that disruption of vestibular inputs to the cortex can powerfully disorganize higher-level frontal functions such as attention. Similarly, noise input due to vestibular lesions can disrupt the intentional or orchestrative functions of the frontal system and, in posterior cortex, memory-retrieval as well, particularly, perhaps, of space-related data.

Following an auto accident, a young woman I know became a vestibular patient. In addition to developing the problems with retrieval and concentration just mentioned, she found that even thinking about spatial relations upset her. If someone asked her the way to the drugstore around the corner, she could remember it but in the process, would become confused and sometimes nauseated.

Lassen et al. (1978), as mentioned earlier, reported that a human subject, asked to imagine a finger-movement without making it, showed a focal increase in cerebral bloodflow in the supplementary motor area. In this vestibular patient, the act of mentally retracing the route from her house to the drugstore may have caused a similar preparatory activation of her motor system, which then acted vertically to trigger noise output from her damaged vestibular nuclei. Fed back, e.g., to the cerebellum, to the brainstem structures shown in Crosby et al. (1962, p. 157, Fig. 132), and to the several higher level structures reached by vestibular projections, this output disturbed her thought-processes and brought on a kind of motion sickness.

In birds and mammals, the developmental sequence is: vestibular, somatosensory, olfactory, auditory and visual. The vestibular system is the first to come on line since it figures in a most basic motor function—"the rapid computation or rapid adjustment of the body's position in space" (Fair, 1987).

As the organism matures, frontal motor-functions, and the related higher-level activities that govern attention and the sequencing or orchestration of behavior as a whole, then develop as rostralmost elements in what is essentially a rostro-caudally organized system, one of whose major brainstem components is the vestibular nuclei. (In chapter 2, above, I suggested the role that vestibular input to primary somatosensory cortex plays in the several-stage development of spatial representation in the parietal lobe.)

Just as the affective or motivational "tags" of long-term memories depend upon diencephalic links that reciprocally connect posterior cortex with the VTA and other lower brainstem structures, "higher" frontal functions remain similarly linked to motor functions and the vestibular nuclei. It is in this fundamental sense—as representing long parallel tracks on which our receptor and effector functions are organized—that the Korsakov's and vestibular syndromes resemble each other. (Shorter parallel tracks, involving the caudate and putamen, are discussed in chapter 6, part 1.)

We are left, however, with this puzzle. In what way are long-term memories in general time-coded? The model I have presented here suggests no solution to that problem. We are able to locate our long-term memories roughly in time, sometimes by a lengthy process of "thinking back." Is the process largely associative, or is association merely a mnemonic aid? Is temporal coding anterior to that, occurring automatically in the interval during which such memories are being consolidated?

Some of the peculiarities of long-term memory, notably its frequent lacunae and disconnectedness, are natural consequences of the structure of long-term memory described in this monograph. We currently have no idea and no evidence as to how our memories may be time-coded, other than by their associational connections.

Memory in Old Age

It seems to be true that we often "lose track of" just when and how important events in our lives took place. We try to "think back" but the associative machinery fails us. As explained in chapter 5, part 3, this may in part be due to the gradual occlusion of older memories, as a result of the probabilistic changes occurring over time in the cortical relay assemblies that figured originally in their formation.

Because, on Blakemore's principle, the response-preferences of assemblies in association cortex continue to be reshaped by the flux of day-to-day sensory experience, some may effectively drop out of the pathways in which they were once consolidated. In proportion as that occurs, a given long-term memory will become "occluded," i.e., resistant to retrieval by associative processes at the cortical level, since the number of interareal paths connecting its core elements to their referents has been reduced. And when retrieved, the memory itself will, for the same reason, be sketchier.

"In the latent state," Luria said, "traces undergo further transformations and sometimes become more generalized and schematic. These changes... begin to distinguish old traces substantially from those recently established" (Luria, 1976, p. 5).

The model of cortical memory functions developed in this monograph predicts these results, just as it predicts that interruption of the white matter connections between distal visual cortex of the lingual and fusiform gyri, and more proximal visual areas is likely to produce deficits in specific but not in generic visual recognition (Damasio, 1985; chapter 2, above). For in this case, both the "core" components and the referents of a visual long-term memory may still exist but have been disconnected.

The vertical organization of memory functions described above would also predict that the "core" components of a visual memory, split off in this way from its referents, might still be activated by the corresponding input. That is, the subject would be unable consciously to identify the input but might show other (e.g., autonomic) signs of recognition.

In two cases of prosopagnosia, Tranel and Damasio reported that the patients, shown pictures of faces unfamiliar to them, mixed in with faces once familiar, gave significant galvanic skin responses to the latter only (Tranel and Damasio, 1985; see Appendix II for details). These findings have interesting implications concerning the nature of "consciousness" which are considered in the final section here.

As we enter old age, the pattern of our remembering begins to change. In this stage of life, in parallel with a decline in our systemic energy, we reportedly show 15–20% losses in frontal and neostriatal neurons, with similar losses or related atrophic changes in hippocampal CA1 units (Squire, 1987, pp. 199–200). These changes may combine to slow down the continual revisions of memory that occur in the parts of cortex that work on Blakemore's principle; and due to cell losses in the hippocampus and frontal cortex, the processes leading to LTM-formation may be slowed down as much or more.

In short, as our general activity level falls, our rate of intake and processing of new information tends to fall with it. As that happens, we become less likely and less able to retain information related to our projects of the moment; and having fewer projects, we become more inclined to drift, letting our moods or emotions call up out of memory what they will.

This is essentially the process by which, as we get old, we slip into the past. By degrees, it turns us into compulsive tellers of old (and usually inaccurate) tales. (See chapter 5, part 3, on the errors that creep into memory due to associational changes in the retrieval pathways.) In the extreme, instead of what we distantly remember being reshaped by what we perceive, what we perceive begins to be reshaped by what we distantly remember.

Thus the very new becomes increasingly difficult for us to "see" or understand, unless it can be recast in more familiar terms. And because it is (as Luria suggests) a *re*-membering, our recall of earlier times has perhaps a core of literal truth, corresponding to the "core" structures of those older memories. *Essentially* what happened is recalled; many of the details may be fictional, and many others lost.

In short, old age may have more to tell us about the structure of our memories than we have wanted or, until now, been able to see.

Chapter 8

Evolutionary Origins of Vertebrate Memory Organization

Part1. The Evolutionary Background of Vertebrate Cortical Organization. The Amphibian Forebrain. The Amygdala and Memory

Abstract. Phylogenetic precursors of the mammalian amygdala and hippocampus. Comparative anatomy of these structures in amphibians; their relation to olfaction, and of the latter to LTM or specific recall in higher vertebrates. Structure of the "extended amygdala." Evolution of "prospective" (action-related) functions of the amygdala and "retrospective" (LTM-related) functions of the hippocampus.

At the end of chapter 5, part 3, I suggested that the mechanisms underlying long-term declarative memory may have resulted from an evolutionary elaboration of olfaction-related functions of the hippocampus in premammals. In this section, I will examine the functional relations between the hippocampus and amygdala, as these have evolved from their primordia in the brain of amphibians.

Originally both structures may have had memory functions related to olfaction, those of the amygdala involving pyriform pallium, those of the hippocampus involving dorsal "general" pallium (Fig. 8-1). In what follows, I will attempt to show some of the transformations these systems have undergone in phylogeny, and what these imply for the present role of the amygdala in memory formation.

The Amphibian Forebrain and Olfaction

In higher vertebrates both the hippocampus and the amygdala receive olfactory input, that to the amygdala being described in the older literature as the more direct (Pribam and Kruger, 1954). As mentioned in chapter 6, part 2, the central amygdaloid nucleus is "a continuation of the lateral olfactory nucleus (nucleus of the stria terminalis)" (Sarnat and Netsky, 1981), which in turn is a major pathway from the amygdala to the hypothalamus.

Nauta questions whether this olfactory connection to the amygdala exists in man. He states that "the primary olfactory cortex (i.e., the pyriform or prepyriform) projects to the amygdala, in large part to a particular cell group, the lateral nucleus..." (Nauta and Feirtag, 1986).

The latter are presumably structures adjacent to, or part of, the basolateral nuclei, which Sarnat and Netsky (loc. cit.) characterize as "phylogenetically progressive" and Heimer et al. (1989) distinguish from the more basal and massively acting "extended amygdala" (discussed below). In the rat (Beaulieu et al., 1988) the basolateral nuclei and pyriform cortex reportedly figure in the development of a "conditioned odor preference."

The hippocampus, by contrast, was defined by Pribram and Kruger (1954) as part of the tertiary olfactory system that is "only remotely related... to the olfactory bulb." According to Nauta (loc. cit.), that is not the case: "The primary olfactory cortex projects to the entorhinal area" and thus, by one synapse more, or probably two, to the hippocampus.

Figs. 8-1 and 8-2, from Herrick (1948), suggest the phylogeny of some of these relations. Note that the amygdala, pyriform pallium and the "somatic striatum" are closely adjacent on the lateral surface, whereas the septum and hippocampal pallium lie medially and include an area Herrick calls "hippocampal dorsal association" pallium (Fig. 8-1).

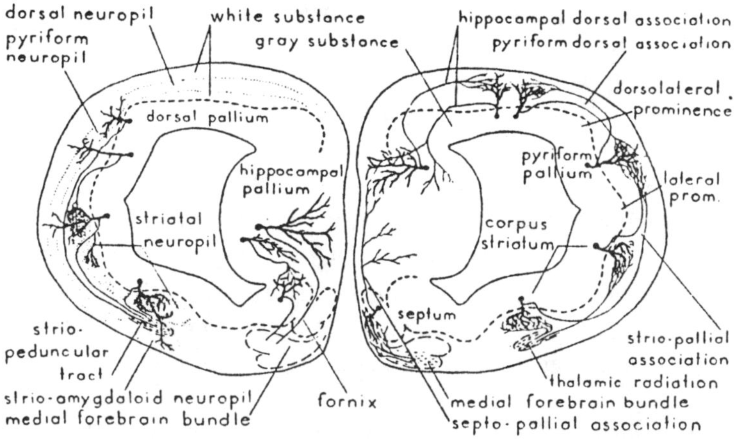

Figure 8-1. From Herrick's Figure 113 (Herrick, 1948, p. 388), showing "the relations of projection and association fibers in the cerebral hemispheres of Necturus, as seen in transverse section a short distance rostrally of the lamina terminalis." Reprinted with permission of The University of Chicago Press, from Herrick CJ (1948): *The Brain of the Tiger Salamander.*

The latter, which appears to end at the "dorsolateral prominence," receives some fibers from the pyriform pallium laterally and the hip-

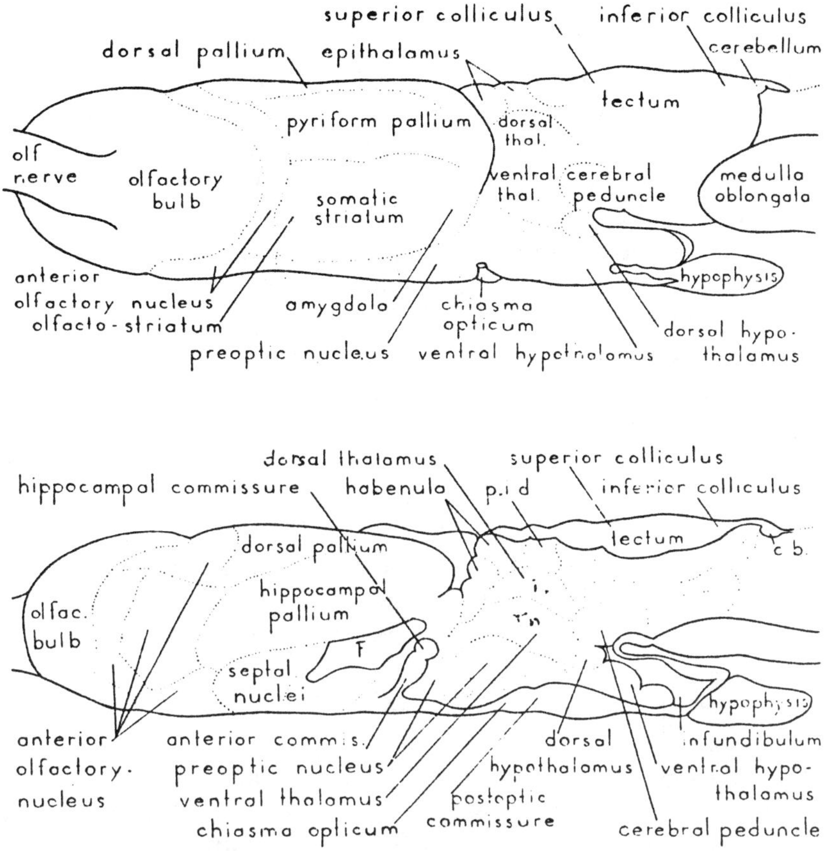

Figure 8-2. From Herrick's Figures 86 B and C (Herrick, 1948, p. 369) showing "the lateral aspect of the cerebrum of Necturus." Reprinted with permission of The University of Chicago Press, from Herrick CJ (1948): *The Brain of the Tiger Salamander.*

pocampal pallium medially. In the course of vertebrate evolution, as the dorsal pallium exploded and became invaded by the specific thalamocortical projection system, the parts of it related to the hippocampus became association cortex.

Thus behind the central sulcus are three islands—the specific sensory and parasensory receiving areas—surrounded by an ever-enlarging sea of association cortex whose far shores, or more distal parts, remain closely connected to the dentate gyrus and hippocampus. It is to these parts of the brain that we seem to owe our capacity to accumulate long-term ("declarative," "episodic," consciously retrievable) memories.

By contrast, the amphibian pyriform pallium, if it ever had as extensive connections as those of the primordial hippocampus with the dorsal pallium, seems to have lost them. For in the mammalian, and especially the primate, brain, it is rolled under and more or less out of the cortical mainstream. It is not a thalamic projection area, but projects *to* the thalamus (as mentioned in chapter 6, part 2). And it is the one part of the brain in which the amygdala appears to figure directly in memory formation (see e.g., Beaulieu et al., 1988).

These pyriform projections, being to the medial magnocellular part of dorsomedialis, probably access orbital cortex, which in turn has direct hypothalamic connections. As a set, these connections suggest a primitive autonomic adjustment system that has survived even in organisms in which olfaction is no longer the dominant modality. A similar survival, reflecting the phylogenetic connection between olfaction and sexual development, is illustrated by Kallman's syndrome (see below).

The structural relations shown here in Figs. 8-1 and 8-2 are given in more detail in Herrick's Figures 97–100 (1948, pp. 376–377). These show the fornix in the medial wall of the ventricle, and (lateral to it) the medial and lateral septal nuclei, suggesting two-way hippocampal-septal connections such as exist in mammals. Ventral to the septal nuclei (Fig. 98) are the nucleus accumbens, and the adjacent bed nucleus of the medial forebrain bundle, which lies in the floor of the ventricle on the medial side.

In the lateral wall of the ventricle, the rhinal sulcus marks the boundary of the dorsolateral prominence shown in the first figure here—a division suggesting some functional segregation of olfactory functions of the pyriform pallium from the (perhaps more general associative) functions of dorsal pallium and medial *primordium hippocampi.* It is this functional segregation that we see carried to an extreme in the primate brain.

The rhinal sulcus lies dorsal to the entorhinal sulcus, which in turn lies dorsal to the densely innervated striatum. In a similar position, putting it, in Herrick's Fig. 97, in the floor of the ventricle, is the amygdala.

The amygdala and n. accumbens thus have similar basal positions in relation, respectively, to the lateral pyriform pallium and to the medial hippocampal pallium. In higher forms, the amygdala (which was once classified as part of the basal ganglia) has indirect motor functions, e.g., in startle responses (Hitchcock et al., 1988), as well as major visceromotor functions by way of the hypothalamus.

N. accumbens also has indirect motor functions in higher vertebrates. For example, 6-OH-dopamine lesions to this nucleus produce a "profound disorder" in food-hoarding behavior in the rat (Iversen, 1986). The medial forebrain bundle (MFB), which is part of the medial septal-hippocampal system in amphibians, forms a major bidirectional link with the ventral tegmental area in mammals.

From the self-stimulation studies of Olds (1975) and others, the MFB is known to mediate intense "reward" responses. Rage and relief of pain have been reported in humans (Sweet, 1980), and pleasure responses have been obtained in psychotics (Heath, 1964) following stimulation in the septal area. Lilly (1960) found that monkeys, with electrodes in the septal nuclei, exhibited penile erection and would self-stimulate, without sleep, for periods of up to 48 hours.

Finally, neurons that produce LHRH (luteinizing hormone-releasing hormone) are found in the septal-preoptic area and hypothalamus, and in the nervus terminalis, which is part of the accessory olfactory system. The latter has been shown to figure in memory formation in female mice, specifically for the odor of a male recently mated with (Brennan et al., 1990). The memories formed last 30–50 days, depend upon noradrenergic input from the locus coeruleus, and can be disorganized (made less specific) by bilateral infusion of bicuculline into the accessory olfactory bulb. In contrast to the main olfactory tract, the accessory tract accesses the hypothalamus by way of the corticomedial nuclei of the amygdala (see these authors' Figure 2).

Neurons releasing LHRH have been shown to originate in the medial olfactory placode in fetal mice, and to migrate to their final loci in the hypothalamus during development (Schwanzel-Fukuda and Pfaff, 1989). The authors suggest that it is the failure of this migration to occur that results in Kallman's syndrome in adult human males—anosmia and hypogonadism.

These findings, taken together with Iversen's concerning the role of n. accumbens in food-hoarding behavior, suggest that evolutionary descendants of the medial septal-hippocampal-n. accumbens system are primarily connected with alimentary and reproductive functions.

By contrast, those descended from the amphibian "strio-amygdaloid neuropil" (Fig. 8-1) may be primarily concerned with adjustments related to direct action of several kinds; that is, not only specific actions related to food-getting and sex, but massive general purpose actions mediated by intense nociception or fear. The traces of this ancient functional dichotomy between the medial and lateral divisions of the amphibian brain can perhaps still be read in the neuroanatomy of the mammalian olfactory tract, which is usually shown as split into the lateral stria, projecting to the amygdala, and the medial stria, projecting to the septum.

The relation of the amphibian strio-amygdaloid system to the type of general purpose or (if need be) massive action described above seems to be reflected in amygdaloid functions in higher vertebrates. Stimulation of the amygdala, in particular the centromedial nuclei, can produce intense fear or startle responses, as well as opioid-mediated rebound responses, resulting in analgesia (e.g., in humans, suffering from intractable pain; Sweet, 1980). In the monkey, opiate receptor concentrations are 3 to 5 times higher in the amygdala than in the hippocampus or frontal pole (Pert, 1980).

A part of the amygdala's capacity to elicit massive results is its capacity to produce massive inaction. If we assume that fear, like our other emotions, has brainstem origins, it follows that the amygdala, because of its extensive inputs, e.g., from the cortex (Price, 1987), may be ideally situated to "decide" what a given fear-producing input calls for. It may act as switching center through which data concurrently activated in the cortical sensory or memory systems can tip the balance of its output—in this case, in favor of "run," "fight," "freeze," or simply "do nothing."

Thus the amygdala (like its complement, frontal association cortex) is an amplifier of basal states such as fear, but because its facilitatory and inhibitory outputs are similarly massive, it can be switched to produce the alternatives just named. Nor is it paradoxical that a system capable of producing intense fear when stimulated can also, when removed, leave the organism at the mercy of its fears (as in the case of the green monkey, described below). For one of its functions is to cancel fear responses if memory-data, input from the cortex, indicate that the triggering sensory stimulus no longer calls for emergency action. Hence its role in "habituation."

In amphibians, the principal forebrain motor outflows appear to arise in the lateral part of the hemisphere—the primitive striatum and amygdala —and to pass by way of the habenulo-interpeduncular system. These relationships survive as the olfacto-habenulo-interpeduncular system in mammals. It is regarded as an ancient accessory motor system. And as mentioned earlier, damage to the interpeduncular nucleus was reported by Thompson (1960) to block the execution of conditioned avoidance responses in the rat, though not the animal's response (squealing) to presentation of the conditioned stimulus.

To repeat: what we may be seeing, in these neuroanatomic data from Herrick, are the outlines of two major systems in the forebrain that are functionally divergent in the sense that one—the system including the primordial hippocampus, septal nuclei, n. accumbens and the medial forebrain bundle—is already more distantly, or one might say, more conditionally related to effector functions than the other.

The other system, based upon the primordial striatum and the amygdala, is one that, in higher forms, seems more consistently related to fear and readiness responses, and so to the powerful activation and powerful inhibitory controls that effective action often requires.

In vertebrate evolution, it is the widening of this functional divergence through the expansion of its neuroanatomic substrates that has led to the fronto-posterior organization of cortical functions that we find in ourselves today.

Because the system involving posterior cortex and the hippocampus is implicated in long-term declarative memory formation, I suggested that it had primarily "retrospective" functions. In contrast, I suggested that the amygdala and frontal cortex were anticipatory or action-related systems, and therefore had primarily "prospective" functions.

Memory Functions and the Amygdala

The amygdala has often been proposed to play a major role in memory formation—one that has been linked to activity in the catecholaminergic systems (Squire, loc. cit., p. 51). It almost certainly figures in the "coactivation" mechanisms outlined in chapter 4, part 3, above. Its "prospective" functions are perhaps reflected in Figure 3-2 above, showing the order in which the amygdala and hippocampus are likely to be accessed by visual data during their receipt in occipital cortex.

While visual inputs via this route evidently reach the amygdala first, it is also obvious that the amygdala and the hippocampal-FD system must work interactively rather than on parallel but separate tracks. One obvious outcome of this synergy is that it would enable long-term memory to modulate incipient behavior, including incipient drive- or feeling-states, in ways consistent with the organism's recalled experience. The other is that the amygdala, because of its apparent capacity to amplify certain affective or drive states, is likely to contribute substantially to the "coactivation" leading to LTM-formation. This is perhaps particularly apt to be the case under conditions of unusual or fearlike stress—conditions demanding action.

The amygdala has cross-modal "hold" functions that the hippocampus reportedly does not. (Murray and Mishkin, 1985.) These clearly are action related and parallel or (via DM) augment those of frontal cortex (chapter 6, part 2). They amount to an attentional mechanism, pinpointing sensory cues the organism may need in order to decide what it will do, or not do, next. Figure 8-3, from Price (1987), suggests the very extensive circuitry that makes these several functions of the amygdala possible.

It is probably significant, in this connection, that the most nearly corticated or phylogenetically "progressive" part of the amygdala, the basolateral group of nuclei, is "developed only in mammals" (Sarnat and Netsky, 1981). It "is especially well developed in the human brain... [where it] shares many features with the cerebral cortex.... As in the rest of cerebral cortex, one of [its] main subcortical outputs is directed towards the basal ganglia" (Heimer et al., 1989).

Along with the thalamic midline and intralaminar nuclei, the basolateral nuclei project to the anterior cingulate (in contrast to the posterior cingulate, which receives projections from the anterior thalamic nn. and the subicular area. Vogt Rosene and Pandya, 1979). According to Sarnat and Netsky (1981) "the basolateral amygdaloid nucleus has a high concentration of acetylcholinesterase activity, distinguishing it from other parts of the amygdala." As in neocortex, the outer or corticobasolateral portion is parvocellular, the inner portion, magnocellular. These authors add that both "undergo constant phylogenetic development."

Finally, they note that this part of the amygdala is "continuous with the overlying hippocampus." However, the amount and direction of activity across this interface are I believe not yet known. The amygdala has extensive connections with temporal cortex (Fig. 8-3), including the temporal pole. It is these connections that may account for the "experiental" responses that have been reported to result from stimulation of the amygdala in humans (Squire, 1987).

Heimer et al. (loc. cit.) divide the amygdala essentially into three nuclear groups, the "olfactory amygdala" reached by the lateral stria, and the centromedial and basolateral nuclei. In mammalian evolution, the centromedial group "remains unchanged except for reduction of the nucleus of the lateral olfactory tract" (Sarnat and Netsky, loc. cit.).

The centromedial nuclear group does not resemble the cortex in structure or have significant dorsal striatal projections. These nuclei form a large unit, extending through the sublenticular part of the substantia innominata into the bed nucleus of the stria terminalis. It is this group of structures that Heimer et al. (loc.cit.) call "the extended amygdala."

It is "closely connected with autonomic and neuroendocrine nuclei of the hypothalamus.... If one considers that elements of the medial [n.] accumbens and olfactory tubercle [or anterior perforated substance] may be part of this continuum, these structures, together with the bed nucleus of the stria terminalis, sublenticular substantia innominata and centromedial amygdala, represent a formidable basal forebrain unit, and the third largest structure in the forebrain, after the cortex and striatopallidal system." It is this complex of structures that gives the amygdala its formidable power to modulate behavior, including its central states of origin, in the ways described above.

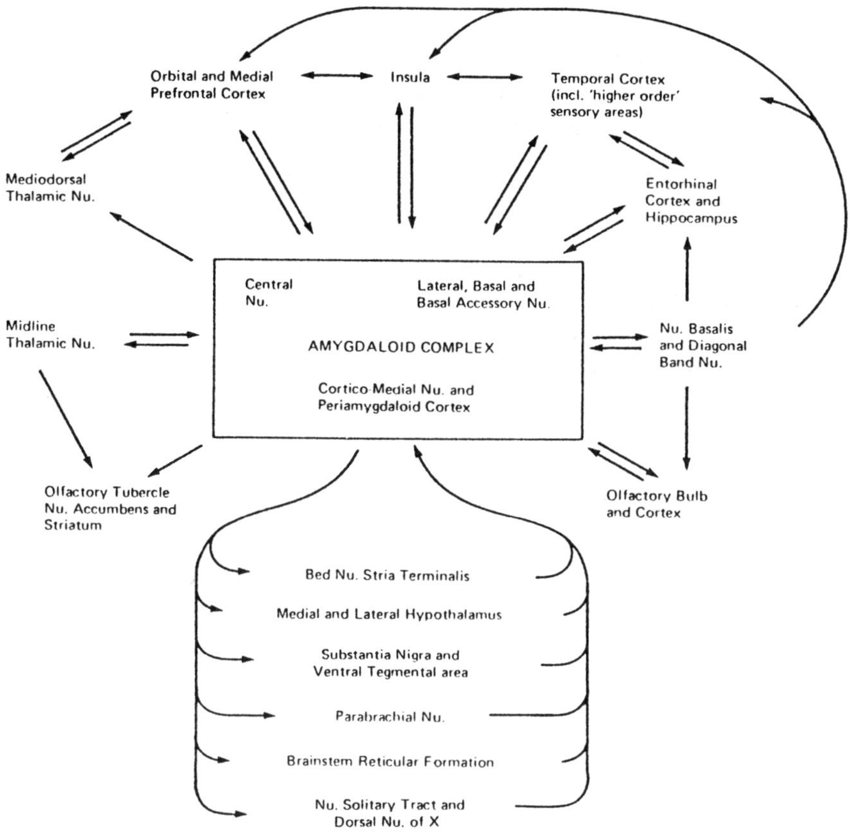

Figure 8-3. Schematic, from Price (1987), showing connections of the amygdaloid complex with other parts of the brain. Price notes that "projections to and from the amygdala may be related to one or several nuclei within the complex." Reprinted with permission of Birkhäuser Boston, from Price JL (1987): Amygdaloid Complex. In Vol. 1, *Encyclopedia of Neuroscience*, G. Adelman, ed.

Its inputs come "from the limbic lobe, orbitofrontal and insular cortex, and via massive projections from the basolateral amygdaloid group.... Another unique feature of the extended amygdala is its direct output to neuroendocrine, visceromotor and somatomotor centers, including the cervical part of the spinal cord" (Heimer et al., loc. cit.).

In the caption to their Figure 1 (here Fig. 8-4), the authors note that the use of histochemical markers shows that the main parts of the substantia innominata ventral to the anterior commissure "are in all likelihood occupied by ventral extensions of the basal ganglia—that is, ventral striatum and ventral pallidum." This is the part of the basal ganglia that

Nauta, among others, has suggested should be designated "limbic striatum" (Nauta and Feirtag, 1986). Its ancestor would be what Herrick (1948) called the "strio-amygdaloid" complex in amphibians.

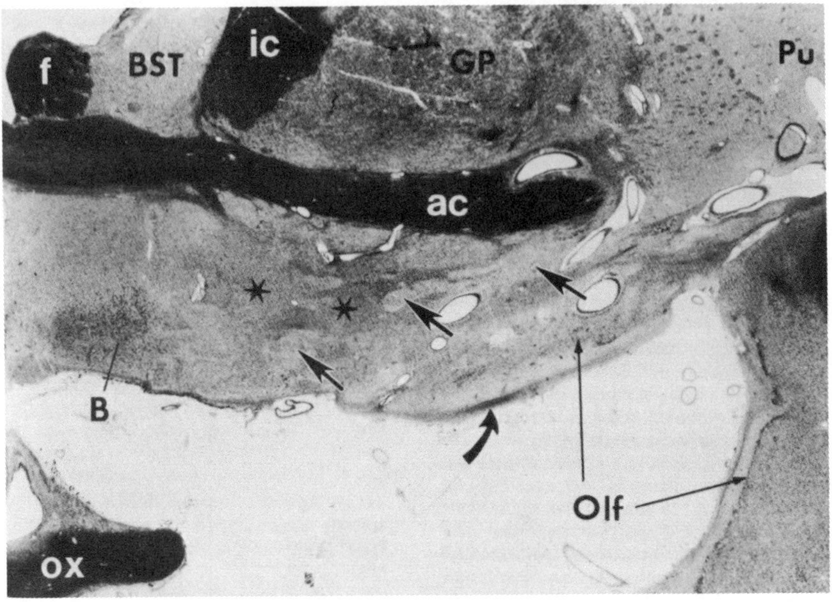

Figure 8-4. Photomicrograph, showing a transverse section through the human basal forebrain, stained with the Klüver-Barrera method for cell bodies and myelinated fibers. The authors note that "the extensive region underneath the anterior commissure is generally referred to as the subcommissural substantia innominata.... The use of histochemical markers has shown that [its] main part... is in all likelihood occupied by ventral extensions of the basal ganglia—that is, ventral striatum (large arrows) and ventral pallidum (stars). Another and more superficial part of the subcommissural area at this level apparently belongs to the olfactory system.... Abbreviations: ac = anterior commissure; B = basal nucleus of Meynert; BST = bed nucleus of the stria terminalis; f = fornix; GP = globus pallidus; ic = internal capsule; Olf = olfactory cortex; ox = optic chiasm; Pu = putamen." Reprinted with permission of Birkhäuser Boston, from Heimer LH, Alheid GF, and Zaborsky L (1989): Basal Forebrain and Substantia Innominata. In *Neuroscience Year*, Supplement 1 to the *Encyclopedia of Neuroscience*, G. Adelman, ed., pp. 21–24.

Below the anterior commissure, in close proximity to the ventral striatum (Fig. 8-4) is the nucleus basalis of Meynert, which, as noted earlier, is a major source of cholinergic projections to the cortex. It shows cell losses of up to 75% in patients with Alzheimer's disease (Coyle et al., 1985), with accompanying losses in the hippocampus. Interestingly, these authors report that excitotoxic lesions to the ventral globus pallidus (but not to the thalamus, dorsal GP or zona incerta) "resulted in selective cholinergic deficits that mimicked those reported in Alzheimer's disease."

Putting all this together, we may conclude that the "extended amygdala" is a powerful and relatively ancient system that mediates the often global, and often very rapid, physiological changes preliminary to overt action (or inaction, as the case may be).

Likewise, the cross-modal "hold" functions reported for the amygdala by Murray and Mishkin (1985) may entail activation (via the extended amygdala) of the (topographically projecting) nucleus basalis. The latter inputs may then act synergically with noradrenergic input from the locus coeruleus to set up the conditions for memory formation in the cortex, as described here in chapter 4, part 3.

Other changes mediated by the amygdala include appropriate shifts in emotional and visceral tone, in neuroendocrine output via the hypothalamus, and in the type and intensity of related drive-states, via brainstem feedback. I have called these functions "prospective" because that is what they are—an immediate adjustment system, bringing about global changes in response to intero- or exteroceptive inputs signalling that some new behavioral course is needed.

In this system, the basolateral amygdaloid nuclei, which are more "cortical-like," more related to the dorsal striatum and, by proximity, to the hippocampus, may have an important informative function with respect to the phylogenetically older and far larger "extended amygdala."

For example, in the rabbit, the basolateral nuclei and two of the areas to which they project (DM and the anterior cingulate) figure in discriminative avoidance learning (Maren et al. 1989). The authors add that "plasticity in the posterior cingulate cortex and anteroventral thalamic nucleus develops more slowly and is more persistent," an observation that is significant since these structures form part of the memory system involving the hippocampus, as described in chapter 7. Indeed, this study suggests, quite specifically, the pathways by which the amygdala and hippocampus collaborate during learning.

In contrast to the basolateral nuclei, the centromedial nuclei have more diffuse or widely distributed outputs, and more basal or elementary functions. Their ability to drive or amplify primitive responses is illustrated by the fact that lesions to the central nucleus are reported to block shock-induced sensitization of acoustic startle responses, apparently in the rat (Hitchcock et al., 1988).

In rabbits, the central nucleus figures in "classically conditioned heart rate responses to acoustic stimuli" and projects (mostly ipsilaterally) to the substantia innominata, thalamic midline nuclei, the ventromedial and paraventricular areas of the hypothalamus, and several brainstem areas including the substantia nigra, periaqueductal gray, locus coeruleus and the nucleus of the solitary tract (Gentile et al., 1988).

Pribram (1989) states that the "amygdala deals with habituation and familiarization." Both are processes in which responses of the arousal or "novelty" type decline as a function of experience. That is, as the corresponding inputs come to be experienced as "nonsignificant," the primary reticular arousal response tends to be damped by inhibitory feedback.

That the amygdala is an important source of such feedback is suggested by the fact (reported by Pribram and co-workers) that in the macaque, bilateral amygdalectomy causes failure of novelty responses to habituate. This might be interpreted as a release phenomenon, resulting from interruption of circuits whereby memory data represented, e.g., in the temporal lobe, can no longer act via the amygdala to suppress arousal responses to stimuli the organism has learned to recognize as not adaptively significant.

(Sokolov, 1960, suggested that memory might cause an inhibitory foreshortening of "orienting" responses. That is, he proposed that "recognition" of an input resulted in cortical output that suppressed arousal. One of the pathways responsible may be the one just described.)

Stimulation of the amygdala in waking humans has been reported to produce a variety of effects, including fear, sensations of visceral tension or churning, and states of euphoria or analgesia. In the monkey, as mentioned above, opiate receptor binding is 3 to 5 times greater in the amygdala than in the hippocampus or frontal pole, and 1 to 2 orders of magnitude greater than in the inferior temporal gyrus or occipital pole (Pert, 1980). Whereas stimulation of the amygdala in monkeys produces a steep rise in plasma corticosteroid levels, hippocampal stimulation produces a slight rise, followed by a prolonged lowering (at 24–48 hours; Mason, 1958).

The amygdala has been found to be essential to survival of the green monkey in the wild. Bilaterally amygdalectomized, the animal reportedly lives in unrelieved fear of its peers, cannot fend for itself, and dies (Nauta and Feirtag, 1986). So far as I know, no such catastrophic collapse of behavior has been reported to follow bilateral hippocampectomy.

In short, the amygdala is as van Hoesen (1985) has described it—a system whose "output is very much directed toward the origin of effector systems that influence motor, endocrine and autonomic areas along the full extent of the cerebral neuraxis." As such, it is an obviously powerful one—far more so than the hippocampus, which, as Swanson (1987) notes, "has only one major subcortical output: to the lateral septal nucleus."

If one compares the projections of the amygdala and of the hippocampus to the hypothalamus, the difference is again striking. As noted in chapter 3, the hippocampus or projections from the subicular area to the hypothalamus may act as a negative feedback on ACTH output (McEwen et al., 1987; cf. the data cited above from Mason, 1958).

In the rat, Swanson and Cowan (1975) report that the hippocampal CA1–CA3 and hilar fields do not project to the hypothalamus but mostly pass via the precommissural fornix to the septum. The fibers in the postcommissural fornix arise in dorsal and ventral parts of the subiculum. Those from the dorsal part, and adjoining pre- and para-subicular areas, project to the mammillary nuclei. Those arising in the ventral subiculum project to the (hypothalamic) ventromedial and arcuate nuclei.

By contrast, the amygdala has two major pathways to the hypothalamus, the ventral amygdalofugal tract, and the longer stria terminalis. The latter projects, in part via the anterior commissure, to the preoptic and lateral nuclei, and to the ventromedial nucleus (described as a "main" target by Sarnat and Netsky, 1981; the connection is reciprocal). The ventromedial nucleus has long been known to have important "stop" functions. Its stimulation can, for example, block feeding initiated by stimulation of the far-lateral hypothalamus.

These differences are further underscored by the fact that anterior cingulate cortex (area 24) receives projections from midline and intralaminar nuclei of the thalamus, and from the basolateral nuclei, whereas posterior cingulate cortex receives projections from the anterior thalamic nuclei and from "the subicular division of the hippocampus" (Vogt et al., 1979).

Although the differences in circuitry just cited represent only a small sample of those shown schematically in Fig. 8-3, or covered in the now vast literature on the limbic system, they suggest a clear basis for the functional distinction made here between these two major limbic structures. These differences, and their functional consequences have, I believe, a long ancestry in vertebrate evolution. Indeed, the two major divisions of the neocortex, the frontal and posterior, appear to have arisen directly out of them, as elaborations of functions of the lateral and medial divisions of the amphibian brain, respectively.

The distinction made here between "prospective" functions of the amygdala and "retrospective" functions of the hippocampus is consistent with Pribram's statement (1989) that, in nonhuman primates, the amygdala is important in response generalization, whereas the hippocampus is important in stimulus generalization.

It is not accidental that in vertebrate evolution, complex memory functions seem to have arisen *pari passu* with the appearance of cortical sheets. Far more closely than the amygdala, the hippocampus has followed that line of evolutionary development. Olds (1974) commented on its "computerlike" structure; and indeed, with the dentate gyrus, it forms a two-layered cortical system. The extension of the influence of that system into neocortex by way of the afferents and efferents described here, was then what gave rise to long-term memory.

The amygdala is largely nuclear in structure. As suggested in chapter 6, part 1, learning in nuclei may of necessity be somewhat primitive. That is, it may tend to reduce to either-or choices or ultimately to switching functions. As noted also, the most nearly corticated parts of the amygdala—the basolateral nuclei—apparently do figure in olfactory learning, in pyriform cortex in the mouse. (See Figs. 8-1 and 8-2 above, suggesting the phylogeny of this relation.) Whether the basolateral nuclei figure directly in learning elsewhere—for instance by way of their projections to the cingulate gyrus—remains to be shown. The data cited from Maren et al. (1989) suggest that their role is a preparatory one, the final stages of memory formation involving the posterior cingulate and its subicular connections.

The gist of the evidence to this point is that the role of the amygdala in long-term learning is secondary to that of the hippocampus. Its primary role, reflected in its global connections and its predominantly nuclear structure, is that of an extremely powerful switching system, whose function is essentially anticipatory—to "decide" from the balance of its inputs, the amount and type of its outputs, or what, behaviorally, is apt to happen next. One consequence of such "decisions" may be long-term memory formation, mediated by the hippocampus and its reciprocal cortical connections.

One may say that "it is as if the amygdala were participating in the brain's appreciation of the world" (Nauta and Feirtag, 1986), but the operative phrase is "as if" since much of that "appreciation" depends upon sensory and memory-data represented elsewhere.

The old controversy as to whether, in "Henry," the deficit in new LTM formation was due to bilateral ablation of the amygdala, seems to have been settled with Squire's patient R.B., who had essentially the same deficit, following bilateral loss of hippocampal function only.

In fact, the amygdala, as a result of the input-output relations reported by Price (1987; Fig. 8-3 above) may figure in attention (as reported by Murray and Mishkin, 1985), and may thus (as suggested by Heimer et al., 1989) contribute to memory formation. The critical distinction that needs to be made is between systems such as the amygdala and the reticular formation, which create the conditions preparatory to long-term memory formation, and systems, such as the FD and hippocampus, that are needed to complete it.

Part 2. LTM and Consciousness. Some Final Questions

Abstract: Implications of this model of memory organization as to the nature and cortical extent of "consciousness." Priming as dependent on memories lying outside the boundaries of consciousness in visual cortex. Motor memories as similarly inaccessible. Effects of frontal and posterior resections compared.

Bottom-up versus top-down behavioral control systems in the CNS. LTM as the basis of, and cortically coextensive with, activities equivalent to consciousness. As frontal effector and posterior receptor cortex have differentiated during evolution to become increasingly distinct systems, the result has increasingly been to permit their functional decoupling. Together with the expansion of LTM, it is this decoupling, e.g., during the performance of habitual or thoroughly learned actions, that has made possible the self-observation and self-knowledge that we identify with consciousness.

What does this inquiry into the nature of cortical memory permit us to conclude about the nature of "consciousness?" The answer is—not much, but what it does permit is at least clear. I will begin with a few essentials.

We know that both in waking and sleeping, the brain shows continuous electrical activity; and that during waking, that activity changes in ways that distinguish waking even from its close relative, the "fast" activity of REM sleep. (Though superficially quite similar, power spectrum analysis has, I believe, shown the two types of EEG to be significantly different.)

During waking, we are said to be conscious if the streams of impressions or perceptions flowing through our minds cause us to act appropriately. We are conscious in the sense of being "tuned in." Intermittently, we are also conscious in the more complex sense of being aware of these internal events as though they were external.

We are able to think about ourselves sitting there thinking, or to think that something we were just thinking was an interesting thought. How, in the brain, is this sort of self-observation possible? While performing some task or watching some event, we seem perfectly capable of switching from a selfless absorption at one moment, to a detached awareness of the whole situation including our own role in it, the next.

There are a number of reasons *why* we might do this. The real question is *how* we can do it. What is it in the brain that makes us able to watch ourselves "going by?" Paul MacLean once remarked that, since we have all the machinery in our heads to think things out and do them, consciousness, in the sense just defined, is really superfluous. "When I am shaving in the morning," he asked, "why do I have to be there?"

The ideas and data presented in this monograph suggest a few clues, particularly as to what one might call the boundaries of consciousness. In part 4 of chapter 4, I reviewed evidence indicating that during LTM formation or subsequent retrieval, playback might not reach as far as prime receiving or immediately surrounding areas—in the visual system, for example, no farther than area 19.

Long-term declarative memories are by definition those that can be retrieved into "consciousness." Is one of the boundaries of our (conscious) visual awareness determined, then, by the fact that playback, e.g., from the "core" components of a given long-term visual memory, stops in area 19, making us unable to retrieve the bits and pieces of visual data that Livingstone and Hubel (1988) tell us are represented in 17–18?

If so, it is perhaps just as well. Moreover, this idea is supported by evidence from old studies showing that electrical stimulation of 17–18 in humans evoked "sparks" or flashes, while stimulation of 19 evoked organized visual images (Fulton, 1951). These findings raise an interesting question, however. Would retrieval of a long-term memory, e.g., of a shower of sparks, involve playback to 17–18—in contrast to retrieval, say, of an organized image of a beautiful woman we once knew? Probably not.

The fact seems to be that our (conscious) visual perception has nothing like the resolving power that Livingstone and Hubel (loc. cit.) report for areas 17–18. So we might have to conclude that visual "consciousness" stops somewhere in proximal association cortex. Until recombination of incoming bits and pieces has gone as far as it has, say, in area 19, we don't see them at all.

In vision, then, there is a preconscious fringe that can be activated by inputs from the periphery and that contributes, as it were invisibly, to what we consciously perceive. The demonstration of that is priming, described in chapter 4, part 4.

The fact that priming cannot occur with nonsense words or geometrically impossible objects implies that priming stimuli, in order to work, must correspond to memories laid down in that preconscious fringe—or outside of (one of) the boundaries of consciousness in posterior cortex. There are perhaps similar boundaries in the auditory and somatosensory systems; and in chapter 2, I suggested that the recognition of phonemes was favored by the relatively poor resolution resulting from the generalization or integration-by-pooling of the corresponding inputs over time.

To put this point another way; it is perhaps adaptively more important for the organism to recognize a given input under a wide range of transforms than it is for it to recall a particular input in great detail. Hence a good deal of detail is excluded, both from direct perception and from memory; hence the boundaries of consciousness in posterior cortex.

Motor memories are not part of this system since they cannot be retrieved directly into consciousness. Or as Squire says: "They are contained within learned skills or modifiable cognitive operations" (1987; p. 152). The demonstration of that was Milner's Henry, who had learned a motor task postoperatively but could recall neither learning it nor the fact that he knew it.

This suggests that mnemonically, frontal association cortex is "silent," as much of it is to electrical stimulation—in contrast to visual area 19, which may figure, for instance, in our conscious perception of visual inputs, and is not "silent" to electrical stimulation.

Likewise, patients who have had extensive frontal lobotomies or topectomies retain consciousness. By contrast, the few patients who have had bilateral temporal ablations (Terzian and Ore, 1955; Burckhardt, 1890–91) have been intellectually devastated, in the same way that Alzheimer patients are devastated by the progressive loss of both short- and long-term memory. In frontal patients, consciousness is altered; in Alzheimer's, it gradually disappears.

So, very roughly, we might have to conclude that the substrate of "consciousness" is memory—specifically as represented in posterior cortex. While much of our conscious thinking may be prompted by inputs from the brainstem, amygdala and frontal system demanding action, what we then *do* tends to be in-formed [*sic*] by what we consciously know.

Conversely, in the course of conscious thought, we may arrive at conclusions that, in effect, reverse these relationships. That is, conscious thought may end up activating the brainstem, amygdala and frontal system, preparing us to do something we have consciously (and perhaps calmly and rationally) concluded may be necessary. It is then this mobilization of the brainstem, amygdala and frontal system that hardens our resolve into the decision to act, partly by the changes it induces in our motivational or emotional tone. ("Drive" takes over; the level of arousal rises, attentional foci sharpen, etc.)

In other words, our behavior and its accompanying conscious thought processes can originate in one of two basic ways—from the bottom up or from the top down—from the brainstem and frontal to the posterior cortical system, or the other way around. Decisions that proceed from the top down we feel are "free;" that is, we made them by conscious deliberation.

Those that proceed from the bottom up we feel are not free. We did not choose them. They forced themselves into our conscious mental lives, and perhaps forced our thinking itself in directions we didn't choose to go. (Hence our extreme ambivalence about sexual love which, however intensely rewarding, is anything but a matter of free choice. Much that we do under the pressure of fear or rage comes under the same head—is driven behavior and therefore humiliating.)

If we suppose that posterior cortex is where our conscious awareness arises, even though it has invisible roots elsewhere, this brings up two other critical questions.

The first is what consciousness *is*—what operations it includes, other, say, than free association. Because of the vertical organization of memory functions discussed in chapter 7, does it invariably have a "deep" component, equivalent to a felt sense of self? Subjectively, one might be inclined to say so, yet observation suggests that people differ enormously in their awareness of their own emotions or physiological states.

The question is, in any case, unanswerable since one of the operations we associate with consciousness is thought, and no one working in artificial intelligence, much less in neuroscience, has come up with a satisfactory definition of that.

The second, and perhaps more answerable question, concerns the bottom-up top-down distinction just made. How does it happen that these relations are invertible rather than (as in most organisms) mostly one way? The provisional answer to that (for more detail, see below) may be essentially quantitative. In humans, the sheer size of neocortex relative to the rest of the central nervous system has become so great that it has begun to compete with lower-level systems for control.

If consciousness includes states of awareness that can look back on (retrieve) similar states that went variable lengths of time before, that would make it, in part, a form of short-term memory—a buffer in which recent data or logical operations could be kept available pending further use. (Further use = thought, which we cannot define)

Frontal "hold" functions, e.g., as illustrated by the work of Goldman-Rakic (1986) on mnemonic functions of frontal visual neurons in the macaque, might provide such a buffer, and hippocampal registration another. Both might be involved in the circular postero-frontal interaction proposed to maintain "roll-over vocabulary" (in the model of sentence generation given above in chapter 6).

STM functions generally appear to work on a "chunking" or drain-and-refill principle, in which the time covered by the chunk varies as a correlate of attention span, and the latter varies as a correlate of "interest." Interest can be generated in either direction (top-down or bottom-up) and will be proportioned to the degree of functional unanimity established in either direction. (The less the unanimity the shakier or more temporary the interest.)

It is because we can form long-term memories representing ourselves that the buffer just mentioned can include recent data representing self as an actor distinct from the recently enacted, even if the latter was an internal event or train of thought.

To rephrase a question raised earlier: if posterior cortex and LTM provide the basis for what we call consciousness, how is it that the functional unanimity that we seem to see between higher and lower neuraxial levels in other forms does not exist to the same degree in ourselves? How is it that in humans, bottom-up control can sometimes become top-down; or at other times can lead to an unstable alternance between the two—a type of "conflict" in which functional unanimity sometimes fails and behavior itself tends to fall into confusion?

The answer lies in the functional divergence that, in the previous chapter, I traced from the strio-amygdaloid and septal-hippocampal systems in amphibians to their primate descendants. Reduced to its essentials, what that divergence represented was a steadily widening split between effector and receptor divisions of the forebrain.

One consequence of that split, as I pointed out in part 2 of chapter 6, was a marked difference in the direction taken by the evolution of memory functions in frontal and posterior association cortex. In frontal association areas, the trend was toward a form of association based more upon temporal than spatial or structural features of the recalled data.

Inputs relating to time or timing, and to concurrent drive or affective state reached that cortex, e.g., by way of the amygdala and DM. The sensory data to be correlated with these arrived by relay from posterior cortex and were proposed to be minimal in structure—just sufficiently structured to act as unambiguous cues or "sign stimuli" in a proto-temporal string (potential behavioral sequence).

It is in these senses that frontal memories are combinatorial and algorithmic, or from a representational standpoint, almost contentless. For the same reason they cannot, of themselves support "consciousness" which *is* representational and therefore a direct outgrowth of posterior cortical memory functions.

Finally, it was the evolutionary extension of hippocampal functions, via the subiculum and entorhinalis, into distal posterior association cortex, and their further extension by the mechanisms of circular interaction and consolidation, so as to include more proximal association areas, that gave rise to long-term memory (chapter 3, part 4 of chapter 4, part 3 of chapter 5.)

As a system, LTM has the flaws we are all familiar with—flaws that are traceable to the fact that in the course of vertebrate evolution, the hippocampus has been "co-opted" to mediate the formation of specific memories in a cortical system primarily organized to support generic memory functions (chapter 2; part 3 of chapter 5; the term "co-opted" is from Dumont and Robertson, 1986).

It is this interweaving of very old with phylogenetically new systems in the cortex that has produced the odd interweaving of these two basic types of memory. The result, in memory as we experience it, is an odd mixture of the fixed and specific with the generalized and probabilistically fluid—a mixture that (as I have tried to show) accounts for the basic features of long-term memory described by Luria (1976), as well as some not included in that description (chapter 7).

Though an evolutionary improvisation, LTM is nevertheless a powerful and indispensable faculty. Without it, our grasp and perspective would be enormously diminished, and thought (whatever thought is) would have very little to work on.

For these reasons, although I can define neither consciousness nor thought with any precision, I would suggest that both have more to do with the hippocampus, posterior cortex and receptor functions than with the amygdala, frontal cortex and effector functions. There is one other point.

Above, I distinguished top-down from bottom-up relations on the neuraxis. The distinction should be extended to include fronto-posterior versus postero-frontal dominance relations, for in these we have a clue to consciousness understood as the capacity for self-awareness. The clue consists in the fact that when we are performing some perfectly routine task we can, and often do, let our minds wander.

What this says is that though posterior cortex may be necessary while we are acquiring a new skill (because we have to think about what we are doing, observe and consciously correct our actions), it is virtually freed when the same skill has become habitual. While learning the skill we haven't time (really the computer space) to be self-conscious about it—or if we *take* the time, may spoil our performance. Once we've mastered the skill, however, we *can* be self-conscious about it, because some of the computer space (posterior cortex) is now minimally involved in the task itself. This brings in the quantitative factor mentioned earlier—the sheer extent of posterior cortex, that enables its activity to develop a qualified independence and qualified powers of its own.

It is this disinvolvement that enables us to "watch ourselves going by." Stated another way, it is the expansion and gradual separation of cortical receptor from cortical effector functions that has led to the gradual emergence of the *Doppelgänger* we call consciousness—the Watcher, the self rooted in retrospection and therefore in an historical awareness of the present.

In fact much of the cortex (as can be seen in the EEG in conditioning studies) may be activated during the early stages of motor learning, whereas in thoroughly learned behavior, involvement of association cortex tends to become minimal (Pribram, 1984). It is the fact that posterior cortical activity is detachable in this functional sense that provides the basis for consciousness.

As an observer system continuously in receipt of inputs from the outside world but only intermittently involved in the development of new responses or nascent habit, it acts as a bridge, linking the specific occasions of our lives into a more comprehensive whole in memory. It is this surplus of central nervous activity, over and above that needed for day-to-day survival, that has led to an awareness of our own continuing existence, and of the world's, apart from what we may be doing in it.

Because of this proto-objectivity, this qualified freedom from control by the "prospective" systems demanding action, we can accumulate long-term knowledge of ourselves, just as we can of things external to us. With due allowance for individual variation (and it is apparently enormous), the freedom to watch ourselves going by entails some ability to learn about ourselves in the process—to develop a reflective awareness of who we are and what we add up to.

The thought processes that call on LTM can thus include ourselves as simply one item of memory among others (though usually of more interest than most). The detachability of posterior cortical processes from others such as those involved in habitual behavior, means that we can think about one thing while doing another. Thought about ourselves is simply a special case of this more general one.

It is exactly because of this detachability of our conscious thought processes from the imperatives of adaptation—i.e., this qualified freedom of posterior cortical activity from brainstem control—that the "functional unanimity" of the whole system can break down. With man, the split between effector and receptor divisions of the brain has become so great that the essentially reflective activity we call thought tends to develop a momentum of its own. To the same extent, thought and fear or desire tend to develop competing esthetics, and we tend to become caught (or simply to oscillate) between irreconcilable drives.

The functions of the FD and hippocampus are, as I have said, "retrospective;" that is, they recycle selected parts of recent sensory experience so that these become temporarily or quite lastingly part of consciously available memory. To the extent that the power or accuracy of reflective thought depends upon the data available to it, the mere expansion of the cortical *capacity* for LTM is a potential expansion of the powers of thought itself.

In mankind, this expansion appears to have passed a kind of boundary, beyond which the degrees of freedom of reflective thought have radically increased. And with that began our expulsion from Eden.

For what reflective thought disclosed was an external world not always or even often amenable to our more primitive wishes, and a self whose abilities, impulses or physical appearance were such as we might not have chosen. It introduced into human life a form of inner conflict not evident in other creatures under natural conditions—a contest between bottom-up and top-down processes for control of behavior.

It radically extended what Freud called "reality-principle" processes to include a rational awareness of futurity, leading to the certain knowledge that we too must die. There is little evidence to suggest that even our nearest primate relatives have a similar sense of futurity, let alone of its implications for themselves.

So the answer to Paul MacLean's question as to why he has to be there while shaving, is that, in long-term memory and therefore potentially in consciousness, he is there, like it or not.

This may not seem much of a solution to the problem of consciousness, but at least sheds some light on the functional arrangements in the brain that may make it possible.

Part 3. Questions Raised by this Account of LTM Formation and its Unitary Basis

Abstract. Some experimental questions posed by this theory. Is the fixation stage in LTM formation demonstrable, or even necessary? Does recursive activity at the synapse during LTP form a memory-like link between the potentiated neuron and its target(s) downstream?

Data suggesting two basic stages in the consolidation of long-term memories, and the possible role of GABA autoreceptors in the second stage. Relation of that stage to phase-locking theories of central processing.

The several-stage generic memory system as leading automatically to the perception of analogical resemblances or the use of figurative speech; and at later stages, as providing the basis for abstract concept formation and mathematics. Possible relation of graded slow-potential activity to processes identified with thought. Thought as resembling Penrose's "judgements" in being pre-algorithmic; as a process (like generic memory formation) by which the brain decreases its own entropy.

The model of cortical long-term memory presented here derives from three basic inferences.

The first is that cortical memory functions in general chiefly involve magnocellular pyramidal arrays whose output is either to other cortical areas or to subcortical structures including the cord. Evidence that supports these conclusions is cited in the text (see also Afterword).

The second inference is that long-term memories are established recursively, by playback from the hippocampus, subicular complex and entorhinalis to association areas. Neuroanatomic evidence shows that such recursive pathways exist. That the hippocampus is the key structure in this recursive projection system is shown by the fact that bilateral hippocampectomy in man makes futher additions to LTM impossible.

The third inference is that long-term memories established in this way have a composite structure, in the sense that only parts of such memories are truly long-term or "permanent" (due to changes in the structure of postsynaptic membrane of the neurons involved). These constitute what I have called the core components of LTM.

They may be regarded as anchor points, upon each of which associatively established pathways converge, or from each of which the referents represented in these pathways may be accessed (retrieved) by divergent return pathways. From several lines of evidence, I have inferred that the core components of long-term memories are laid down in those areas of association cortex that are closest, both synaptically and by phylogeny, to the hippocampal-entorhinal system.

The remainder of such memories—their referents or associative structure—consists of components that, because unfixated, are subject to further probabilistic change. Memories of the latter type take shape in response to the day-to-day flux of sensory experience; and in proportion as they do so, act as a filter that channels incoming sense-data according to the experientially established categories with which these show the best fit.

Thus LTM is a composite of indefinitely long-lasting core structures and of referents that are less stable because the corresponding memories tend to show the changes just described.

Two lines of evidence support these inferences. The first is that in the visual system, "generic" recognition is mediated by cortex lying synaptically closer to the primary visual receiving areas than does cortex mediating specific (visual) recognition (Damasio, 1985). Secondly, this clinical finding is supported by data from psychophysical studies showing that generic visual recognition in fact precedes specific.

That LTM, as proposed here, has a composite structure, part fixed, part susceptible to change over time, is supported by psychological studies, in particular those of Luria (1976). What remains to be demonstrated are the mechanics of the "fixation" process that distinguishes the core components of a long-term memory from its more labile referents.

I have supposed that that process may be biochemical, but the evidence to date that neurotrophic factors may act in this way—so as to perpetuate functionally induced changes in postsynaptic membrane—is scanty. An alternative, and more parsimonious, hypothesis, would propose that all memory formation proceeds on Blakemore's principle, but that the rate of change of such memories declines as a function of their distance from prime receiving cortex. This concept (developed in item 7 below) would define core components of long-term memories as those whose probability of further change had become negligible.

As an investigator testing my own theory, I would want to ask or do the following:

(1) Establish the precise distribution, by layer as well as by area, of return fibers from the hippocampus or subiculum or entorhinalis to neocortex, e.g., in the macaque. How many synapses would there be in these pathways? Would there be fewer, say, in projections to the homologue of the fusiform gyrus than in those to area 22?

(2) Could units at the origin of some or many of these return pathways be shown to release growth factors and/or other neuroactive substances specific to those pathways?

(3) If so, could antagonists that blocked postsynaptic receptors for one or more of the foregoing, be shown to block LTM formation? (It is already known that antibiotics that block protein synthesis can have this effect.)

(4) Can neurons proposed to learn probabilistically be demonstrated to show the dual changes in threshold proposed here—lowered thresholds to most probable inputs, raised thresholds to inputs of lower (de facto) probability?

(5) Can these units be shown to have undergone corresponding stuctural changes in postsynaptic membrane?

(6) Might similar changes be shown for postsynaptic membrane in neurons that form part of the "core" structure of long-term memories?

(7) If so, what would differentiate the changes in 6 from those in 5? This really comes down to the question of whether core elements in LTM require special fixation mechanisms at all. For as generalization is proposed here to occur during intracortical relay, core elements, being represented at the far ends of the relay system, are maximally generalized. They would therefore be those most resistant to probabilistic change over time.

This would mean that although far more stable than many of their referents, they would still not be absolutely so. It would mean only that a given long-term memory consists of components of graded probability, such that its most detailed referents would also be those most likely to change or drop out of the memory with the passage of time.

The vertical organization of memory functions discussed in chapter 7 is relevant to this point. We have noted that core components of LTM—those responsible for specific recognition—appear to be represented in distal association cortex. In contrast to the proximal association and prime receiving areas, this cortex has close connections with the limbic system, and thus with the brainstem. These connections provide a set of rostro-caudally running loops over which circular reinforcement, particularly affecting the core components of LTM, could occur.

As in the case of circular hippocampal-cortical interaction, a presumed effect of activity over these loops would be to increase the strength of synaptic connections already in being, in that way increasing the resistance of core elements in LTM to further probabilistic change. Hence, perhaps, the failure of remote memory in Korsakov's syndrome, when some of these loop connections are cut in the mammillary bodies.

An advantage of this hypothesis is that it would obviate the need to postulate special postsynaptic fixation mechanisms to account for the duration of long-term memories. Nor would it substantially alter the account of LTM structure given here. What it would not explain is the fact that LTM formation evidently requires an intact hippocampus whereas retrieval of long-term memories does not. The hippocampus must contribute something to the formation process that makes such memories, at least in part, uniquely independent of dynamic maintenance.

The question then becomes the one raised in part 3 of chapter 5—that of whether the "chemical addressing" of olfactory inputs to the bulb was translated into olfaction-related memories by the *primordium hippocampi*, e.g., in amphibians. A necessary characteristic of such memories is their specificity (for certain odorants). The function of the primitive hippocampus may have been to translate the specificity of certain adaptively significant odors into correspondingly tagged (hence specific) cross-modal memories, by way of projections that biochemically fixated such memories in dorsal pallium. (Such a mechanism, if it exists, should be demonstrable in present-day reptiles or amphibians.)

In other words, the specificity of the input was provided by the olfactory bulb, and that of the input together with its cross-modal associations by the hippocampus

I have supposed that the hippocampus has continued to have essentially the same function in higher vertebrates, the resulting memories being fixated only in part and (except where olfaction is involved) specific only in their patterns.*

(8) To what extent may memory formation involve recursive action or feedback at the synapse, establishing a link of some duration between the pre- and postsynaptic neurons? During the establishment of LTP, might such feedback cause complementary changes in postsynaptic receptor membrane of the potentiated neuron, and in the area of "active zones" (Bailey and Chen, 1983) of the corresponding presynaptic axon terminals?

In fact, Malinow (1991) suggests such a mechanism. Speaking of

---

* Some very old work by Pechtel and Masserman (1954) indicated that monkeys, trained in an aversive response paradigm, showed normal extinction when cues to the response were auditory or visual. When the cue was olfactory, the response was extremely resistant to extinction, and during training tended to generalize to chemically similar odors (possibly because of the involvement of the amygdala in olfactory learning). For some time after apparent extinction of the response, the animals appeared to be disturbed on presentation of the original olfactory stimulus.

LTP, he says: "After its postsynaptic induction a retrograde messenger could be released, affecting Ca^{2+} homeostasis and yielding Ca^{2+} oscillations and increased transmitter release." Thus "synaptic inputs active during the induction of LTP (may) become associated by showing synchronized, oscillating potentiated transmission."

If the mechanism Malinow proposes can be demonstrated experimentally, it would distinguish LTP from sensitization, since in the latter the principal change appears to occur not in receptor membrane but at the axon terminals. At this writing, the consensus appears to be that LTP plays *some* role in memory functions, and debate as to whether LTP is primarily pre- or postsynaptic obviously relates to that question. In vitro experiments involving hippocampal or cortical slices might show that (as Malinow suggests) it is both.

(9) Finally, to what extent may the "coactivation" leading to LTP depend upon the reticular inputs described here (chapter 4, part 3)? To what extent may it involve specific-nonspecific convergence and therefore not necessarily be "associative" in the classical Hebbian sense? That is, might specific input from a single source, when coupled with nonspecific activation of sufficient intensity, be able to induce it?

We stand, I believe, at a critical point in our understanding of the brain. A key to that understanding lies in the circuitry and the unitary changes responsible for the several forms of memory.

Because our knowledge of the *minutiae* of cellular mechanics is now accumulating so rapidly, some of the suggestions made here—for instance as to the mechanisms underlying long-term memory formation—may soon (or already) be out of date. And some of the data cited in this monograph may provide important clues to that process that I have either understated or simply misunderstood.

Let me close with an example. In part 2 of chapter 5, I reviewed the report of Silva et al. (1991), who found that "deep" layer V pyramids of rat somatosensory cortex (in vitro; slices) showed intrinsic 5–12 hz firing patterns.

In a more recent study, Mott and Lewis (1991) reported that in rat hippocampal slices, induction of LTP in dentate gyrus cells by low-frequency repetitive stimulation was facilitated by $GABA_B$ receptors, a finding they note has been replicated for CA1 cells by Davies et al. (1991). The effect was frequency dependent.

If the stimulating pulses were delivered 200 msec apart, the second pulse in the train resulted in a reduced $GABA_A$-mediated inhibitory postsynaptic potential—a reduction that could be reversibly blocked by 2-OH saclofen, a $GABA_B$ receptor antagonist. When ten stimulus pulses were delivered either at 0.5 or at 100 Hz, this reduction was not obtained.

In other words, at 5 Hz—the theta rhythm associated with learning in animal studies—there was a maximal inhibitory suspension of $GABA_A$-mediated inhibition, presumably by way of $GABA_B$ autoreceptors. The result was then a net increase in excitation in the assemblies involved.

The authors note that this $GABA_B$-mediated effect may not only not occur at other stimulus-frequencies, but may also, at some, interfere with LTP-induction. For example, they cite a study showing that LTP induced at high frequencies was enhanced by $GABA_B$ receptor antagonists—suggesting that extremely high levels of arousal may interfere with LTP-induction and so with memory formation.

For the converse to occur—for GABA autoreceptors to contribute maximally to LTP—apparently requires an optimal interstimulus interval. Note also that that interval falls in the lower part of the frequency range reported for the "intrinsic" firing rates of deep layer V pyramids reported by Silva et al. 1991)—i.e., 5 Hz vs. 5–12 Hz.

Taken together, these findings suggest that LTM formation may occur in two basic stages. The first involves rises in reticulocortical input favoring "coactivation" as described in the text here. These will be accompanied by parallel rises in $GABA_A$ mediated inhibition, partially antagonized at higher frequencies by GABA autoreceptors.

The second stage, occurring in what I have called the "downslope" phase of central activation, may then involve longer-lasting circular processes over the pathways initially marked out by coactivation. One of the mechanisms sustaining this circular activity may be the $GABA_B$- or autoreceptor-mediated inhibition of inhibition just described. Called into play as the central nervous activity returns to near-resting-state levels, it may serve to complete the process of consolidation.

Consolidation is known to be time dependent and to occur on a scale of minutes to hours, or even days. It may thus long outlast the states of central arousal in which the whole process of LTM formation is proposed here to originate. It may consequently depend upon the type of low-frequency-mediated LTP reported by Mott and Lewis (1991)—the prolongation of local excitatory activity by $GABA_B$ autoreceptors. This would represent the second or final stage of LTM formation, occurring in the extended aftermath of some precipitating incident.

During it, one might expect to see some phase locking of hippocampal output with intrinsic rhythmic activity of layer V pyramids (Silva et al., 1991) in those parts of the neocortex reached by return fibers from entorhinalis or the subicular complex. Layer V pyramids, it will be recalled, form the basal or final output portion of the two-tiered cortical memory system proposed here.

It is, in short, as we work out the details of findings such as those just cited, that the nature of long-term memory itself may finally become clear—that we can begin to see it not as a patchwork of discrete mechanisms but as a definable, if less than perfect system. Let me conclude by pointing out the features of this model of memory functions that I believe are new.

The concept that in the cortex, generic memories are laid down forward (from prime receiving to association areas) and specific long-term memories backward (by recursive circuits that approach but do not include prime receiving areas) is, in itself, not new. In the case of LTM, it has been latent in the data, at least since the work of Flexner et al. (1963) on consolidation in the mouse, or since Milner's studies of the bilaterally hippocampectomized Henry (Scoville and Milner, 1957; Milner, 1963). What is new is the degree to which the implications of that idea have been worked out here.

That specific long-term memories are partly fixed and partly subject to change over time is something we know from everyday experience. My account of *why* they are, and of the evolutionary makeshifts that may have led to that result, is, I believe, quite new. So is the idea that consciousness is directly linked to the recursive processes that figure in LTM formation or retrieval. Consciousness is thus essentially retrospective—an arrangement for retrieving those specific data in memory most likely to be relevant to our current circumstances or intentions. It enables us, by looking backward, to make informed decisions about how to go forward.

Because those recursive processes reach only so far in the cortex, consciousness itself has cortical boundaries. It is outside of these that the psychological phenomenon known as priming occurs. To date, there has been no clear physiological explanation of priming; nor has anyone, until now, explained why it only occurs with real words or with pictures of geometrically possible objects. (It does so, because these correspond to memories that are preconscious, being laid down in cortex not recursively accessed by the LTM system.)

The account of generic memory functions has important implications for cognitive psychology, since it shows that categorization and abstraction arise automatically in the course of the several-stage cortical relay of primary sense-data. Memory formation of this type occurs primarily in the "forward" mode, and is precognitive in the sense of supplying "consciousness" with many of the forms of our thinking ready-made.

Hence metaphor and metonymy—figurative speech—do not have to be taught. They come naturally to us; and analogical reasoning is a primitive first form of thought. In language, much naming works by analogy, indeed often by associations related to body parts—the "eye" of the storm, the "eyes" through which we put our shoelaces, the last "leg" of a journey. We may reject an argument (perhaps including this one) "on the face of it." And so on.

In other words, the generic memory formation that leads to categorical recognition (of sensory inputs) also leads on to the formation of second-order generic memories representing invariances common to otherwise unrelated memory-data. These memories presumably arise by the same processes of intracortical scansion and probabilistic retention that result in category formation—the difference being that the data extracted in this case are obtained across categories, and represent a later stage in the relay.

It is this system of second-order generic memories that supports the recognition of analogical resemblances, and causes us to use or understand figures of speech almost effortlessly. That the operations leading to first- and second-order generic memories are essentially the same is due to the fact that the layer III-to-III relay system and the structure of six-layered generalized eulaminate (association) cortex are themselves somewhat stereotyped. Each stage or level in the system includes lateral connections equivalent to associations established experientially between memory-data represented at that stage or level. Such connections, e.g., involving the shorter projections arising in layer II, provide the probabilistic structure of memories at that level of generality—for instance the combination of features representing "dog."

Thus each stage in the relay, though having forward and recursive connections (and therefore potential referents at other levels of generality), has an associative structure and a degree of generality peculiar to itself.

So just as repeated exposure to real-world dogs results in "dog," repeated exposure to nearly or perfectly circular forms results in the higher-order generic memory "circularity." Whereas "dog" represents a certain range of general characteristics within a particular category (living things), "circularity" derives from inputs that cross many categorical boundaries, and thus represents a further stage in generalization.

A memory item of this type, as it derives from such a diversity of real-world data, has so many possible referents as effectively to have none. It has begun to be what we call abstract. The striking point here is that no special cortical mechanism or cognitive device is needed to account for this result. It follows simply as a natural consequence of the repeated generalization mediated by the interareal relay system—that is, by the magnocellular pyramidal arrays that are the principal links in that system (see Afterword).

To the extent that that system allows us to accumulate memories of successively higher orders of generality, it automatically provides the basis for abstract thinking. Its ultimate effect is to stock the brain with a set of memory objects so remote from their miscellaneous origins in direct sensory experience that their most clearcut associative connections are potentially with one another. It is because such higher-order memories have many possible but no *necessary* real-world referents that the brain's logic functions are maximally free to operate on them. In turn, it is that freedom from necessary external referents that enables us to construct purely logical, self-referential systems such as mathematics.

As explained in the text, such logic functions may derive, by a similar, several-stage generalization process, from the frontal systems that orchestrate our overt behavior. It is for this reason that we find frontal and posterior association areas connected in a way that systematically reflects the degree to which each is proximally or distally related to primary sense reception (posteriorly) or to final motor outputs (frontally).

That is, the frontal and posterior systems are matched in respect of the form of the memory-data represented in each. Their proximal-to-proximal and distal-to-distal interconnections mean that those parts of frontal cortex mediating short-term (tactical) behavior access more nearly immediate sense-data posteriorly. Those parts mediating longer-term (strategic) behavior access posterior areas in which both higher-order generic memories and the "core" elements of LTM, or specific memory, are represented.

Note that LTM, or depth of recall, is apt to be particularly important in complex planning or strategy. Note also that strategic behavior is likely to arise from persistent strong drives or deep emotions—from sustained rather than purely momentary intentions. The physiological correlate is that distal association cortex, both frontally and posteriorly, is more closely related than proximal, to the limbic system (Graybiel, 1974), and thus to the basal sources of motivation.

The orderly return projections from frontal to posterior association areas likewise tend to guarantee that frontal logic functions of a given order of generality or abstraction will call upon memory-data of the same order. In the case of mathematics, these are the higher-order generic memories mentioned above. It is perhaps only in man that there are generic memory systems, both frontally and posteriorly, which are sufficiently extensive and sufficiently generalized in their contents, to support that sort of "pure" (nonreferential) mental activity.

In mathematics, what we see is the machinery that underlies planning, or the strategic organization of behavior, beginning to operate *in vacuo*—as a self-contained system, running on logic perhaps originally derived from that of overt action but now largely free of the constraints that action itself imposes. It is odder still that a system, freed in this way to operate according to rules generated within itself, should have such remarkable power to define or, by extrapolation, to anticipate the structure of real events.

In sum, what this monograph attempts to show are the several ways in which the generic and specific memory systems provide the raw materials for what we call thought. Nor are all of those materials in fact raw. For it is the cytoarchitecture and circuitry of the cortex that causes sensory data to be generalized and associatively cross-connected as I have described; and it is the resulting system of generic memories that causes us to categorize our sensory inputs, to recognize analogies or to understand figures of speech—all without apparent effort.

As described above, a further stage of the generic memory system may then provide the basis for concept formation and abstract thought; while the LTM system, interwoven with this one, provides much of the data on which these logic functions can then operate. To put this another way: the generic memory system provides the *forms* or (frontally) the sequential structure likely to be taken by thought, whereas LTM or specific recall provides its materials, including background data from the past that generic memory alone cannot supply.

The expansion of these two memory systems, occurring *pari passu* with the evolutionary expansion of the neocortex, appears finally to have resulted in a state change, such that in *homo sapiens* the processes equivalent to thought have tended to become semicontinuous rather than episodic.

In other words, as the base of consciousness in LTM became broad enough to permit ad lib reflection, and as generic memory added the further stages needed to support ad lib logical extrapolation, the result was to make us the only species capable of language, and the only one afflicted with the knowledge that we must die. To suppose that we developed language because we had developed special mental organs for it is contemporary nonsense. We developed it because, on a critical mass principle, the central nervous processes equivalent to thought had become so continuous that they *demanded* some form of organized overt expression.

Other organisms understand the symbolic—that sign A can "mean" event B. We simply systematized that sort of understanding, making it a mode of continuous transaction between individuals. For the same reason, as I have tried to show, language is a form of action, organized in essentially the same way as our actions generally.

(The controversy over whether apes or dolphins can be taught to talk might never have arisen had the original question been put properly. Apes can be taught to signal a variety of things, including curiosity. The real question is can they be taught to *converse*, to wonder aloud and at length about anything? Are their thought processes sufficiently self-propelled to support that kind of activity? The evidence to date suggests not.)

What thought itself is, what neuronal processes it corresponds to, is another question—though on that, I tend to agree with Penrose (1989). The connection-forming or extrapolative activity we call thought may ultimately depend upon the play of slow potentials, and therefore be far subtler and more flexible than the all-or-none activity in which it is likely to eventuate.

Consequently the subtlety of our intentions or ideas may often be lost in the translation. This may even apply as between sectors in the interareal relay system. The relay from one to another being by way of magnocellular bands and/or thalamic association nuclei (and hence all-or-none), there may be some loss of information in transit—an arrangement that, if it is the actual one, would give physiological support to the current notion of intelligence as modular. It would also imply that any form of overt behavior including the symbolic is apt to be cruder than the brain events in which it originates.

If what Penrose calls our "judgements" arise in this way, through new connections established by graded electrical activity, it follows that he may be right. Thought is not algorithmic, but a kind of improvisation in the living present, that makes do with the data available to it. These are mostly, but not entirely, the data of memory.

Memory, as we have seen, may be a matter of neuronal connectivity, established *de novo* in the course of the organism's experience. Thought is perhaps an extension of the same process. That is, it arises out of the brain's free energy, and acts to establish further connections amongst the data of sense and memory—connections frequently related to intended actions (symbolic actions included) and therefore having an important frontal component.

Many of those connections, if not the majority, tend to be transitory. They correspond to what we call passing thoughts or ideas. However, on the same principle of reinforcement that leads to generic or specific memory formation, some thoughts may also become items of recall. That is, the transient connections equivalent to an idea may become more or less lasting, and (like memory in general) shape our future thought processes accordingly. These residues of thought are what Penrose calls algorithms—sets of learned principles, or yesterday's insights hardened into "rules."

One can think of the brain in thermodynamic terms, as a system that draws on the energy supplied it by the vascular bed to support operations that tend to reduce the entropy of its own contents (Fair, 1984). Memory does that, for example by organizing the data of perception into categories, thereby reducing the central nervous energy needed for recognition. It foreshortens "novelty" responses or what Sokolov (1960) called the orienting reflex.

Thought acts similarly, by working over the data of memory and those of primary perception so as to make them more comprehensible. To comprehend means to "hold together"—to grasp a set of things as, in some sense, a system or working whole. Thought is a form of free central nervous energy that connects up the factual data of sense and memory in ways that make them more *available* to us, either as guides to future action or in the more general form known as understanding.

In elementary physics, entropy used to be defined as a measure of the unavailability of a system's energy for useful work. Thought is negentropic—an entropy-reducing process whose most successful forms tend, through repetition, to become items of memory themselves. As such, they are no longer thoughts but formulas—mental mechanisms, functionally more akin to the all-or-none relays of the memory systems than to the graded electrical activity in which thought itself may arise.

Through mimesis (including forced mimesis or education) these formulas then become the intellectual repertoire of a whole generation, and help to blind it to innovations its successors may have in store. However much we may deplore it, this is a limitation we have no choice but to accept. For the brain is, after all, a finite system; and to the extent that the free energy of thought takes algorithmic form or sets into habit, it acts (as do the contents of memory in general) to limit the degrees of freedom of thought itself. Sooner or later, we become prisoners of that process—of original insights, most of which were not even our own.

Afterword

> Evolution does not produce novelties from scratch. It works on what already exists, either transforming a system to give it new functions or combining several systems to produce a more elaborate one.... In short it works like a tinkerer who uses everything at his disposal.... For an engineer, the realization of his task depends on his having the raw materials and the tools that exactly fit his project. The tinkerer...manages with odds and ends.
>
> François Jacob, 1978

The brain is a bewilderingly complex result of this kind of evolutionary make do, containing, as it does, an enormous assortment of centrally-acting substances such as TSH, CCK-8, somatostatin, neurotensin, 8-AVP, substance P, etc., which have portmanteau functions; or of genetically variable receptor systems, such as the "flip" and "flop" versions of the four AMPA receptors discussed in chapter 5, whose functional significance remains to be made clear.

When a phylogenetically new motor system such as the pyramidal is added, it is simply superimposed upon its predecessors. The engineering equivalent would be to improve your car's engine not by replacing it but by adding a second one and tinkering up the two so they worked together. How confusing the results of this type of construction can be is shown by the fact the we still do not fully understand how the skeletal muscle spindle system works (Matthews, 1988).

One reason, as suggested by Henneman (1991), is that the system may make use of built-in errors—essentially, the tendency of branching axons to show transmission failure at some of their branches. Since this error rate may be inversely related to the frequency at which the neuron is driven, it provides a way of grading muscle responses over a wide range of tensions, without need of a complex central switchboard to achieve the same result. (Henneman in fact cites this as a possible example of evolutionary tinkering.)

In a commentary published in *Nature*, Huszagh and Infante (1989) wondered if "the hypothesis in biological sciences" might not be "an endangered species." Theorizing about the brain, however, is extremely difficult because the threads of logical consistency that run through its evolution are easy to lose in the mass of ad hoc arrangements that evolution has also patched together to make the whole thing work. One such thread, as I will try to show, is represented by the organization of cortical memory functions proposed here.

That hypothesis postulates that memory tends in general to be a function of magnocellular output assemblies. In neocortex, these are principally the pyramidal arrays found in lower layer III (which sends output to other areas), and in V (whose main outputs are subcortical). The evolutionary logic that has led to this arrangement is straightforward; to grow, feed itself and reproduce, the organism must first of all survive. To do that, it must be able to act appropriately under a considerable range of conditions, many of them impossible to anticipate in the form of specific innate responses.

Therefore, the primary function of memory is to retain *outputs*, whether these be directly motor (as in learned actions), or indirectly action-related (e.g., as in conditioned stimuli). The essential requirement of the retained data in either case is that they be *available* on short notice—not too fine grained, or needing too much time or energy build-up to come into play.

The generic memory system meets these requirements. As illustrated in the case of phoneme recognition (chapter 2), what it retains is the principal invariants extracted, over time, from a range of never-truly-identical inputs. The details of the latter are supplied by way of parvocellular "processing" assemblies, and magnocellular bands in the same cortex then do the extracting. As mentioned, one of the shortcuts in this system, making generic memories promptly available, may be provided by the thalamic association nuclei (lateralis posterior and the pulvinar). Another system of shortcuts is outlined below.

It is important to see the foregoing not as a special adaptation subserving a particular function (memory), but as one instance of a more general principle of organization. For the magnocellular bands that act as invariance detectors in auditory or visual cortex are, in fact, part of a much longer chain of similar arrays extending from the retina and magnocellular lateral geniculate (or the magnocellular medial geniculate), to motor cortex and the ventral horn.

Throughout, we find the same basic pattern; large-celled, fast, phasically-acting throughput arrays, each juxtaposed with smaller-celled, slower, tonically-acting assemblies having accessory "processing" or modulatory functions. Like layer II in the cortex (Sarnat, 1987), the high-resolution retinal ganglion X-cells (Wright and Ikeda, 1974), or the color-and-spatial-frequency sensitive "parvo" system of Livingstone and Hubel (1988), these "processing" arrays tend to be phylogenetically recent.

The refinements of movement or perception that they permit are likewise slow to mature. Developmentally, axons are the first to come online; dendritic growth and arborization come later. On the same pattern—putting action first—the vestibular system is the earliest to become functional, since awareness of the body's position in space is a precondition of action generally.

Consequently, throughout much of the nervous system, we find parvocellular processing assemblies sandwiched together with magnocellular throughput arrays, in such a way that the latter can retain the resultants of processing by the former and pass them along promptly as needed. For memory to be represented in throughput arrays therefore makes memory, as an influence on behavior, efficient. And since throughput, or ultimately action, is the prime function of the whole system, processing arrays are essentially auxiliaries, arranged so that they can be by-passed as circumstances may require. *This is why the input to all cortical areas is to the midregion—layer IV or lower III. By accessing arrays of both types, it leaves open these alternatives.*

The magnocellular medial geniculate (part 3 of chapter 4, above) provides an example of such by-passing functions. It projects non-tonotopically to all auditory fields and responds not only to auditory but also to somatosensory inputs, and to what the authors describe as "noxious" stimuli (Weinberger et al., 1990). These features suggest that MMG receives input from the reticular formation and acts phasically as one of several triggers capable of setting off a massive behavioral response. In parallel, its diffuse projections to primary auditory cortex may effectively override the more precise tonal analysis normally carried out there.

(Exposed to threatening sounds, or to some combination of sound and "nociceptive" stimuli such as pain, the organism does not need fine discrimination of pitch, timbre, etc, and the auditory system is organized so as to set it aside, in part probably by way of accentuated inhibition.)

In the "magno" visual projection system of Livingstone and Hubel (1988), we find a similar action-related shortcut. For though it has projections that overlap with those of the parvo system in area 18, the authors report that the magno system also projects directly from area 17 to the temporal MT area, and thence to motion-related cortex in the parietal lobe.

(Moving objects or creatures in the organism's visual field are more likely to be adaptively significant than stationary ones. It therefore makes sense that information of this kind, stripped of fine detail including color, should be handled by fast-acting magno units that by-pass some of the usual processing stages in order to access parietal cortex more directly.)

That tonically acting "processing" systems are more energy-intensive, because more continuously active, is shown by the fact that the visual parvo system in area 17 is concentrated in what the authors call "blobs"—patches that stain heavily for the mitochondrial enzyme, cytochrome oxidase; or in 18, in the similarly dark-staining "thin stripes."

(Parvo cells, for example, are reported to respond more slowly than magnos to increases in center-surround brightness differences, but they also saturate at much higher contrast values, indicating greater resolution along this stimulus dimension. Like the greater spatial resolution of the parvo system, this feature comes at a higher metabolic cost, and is hence reflected in differentially high concentrations of cytochrome oxidase found in the "blobs.")

The "thick" stripes in area 18 appear to receive both parvo and magno input, a convergence that perhaps marks the first step in the construction of generic visual memories. Livingstone and Hubel (loc. cit.) appear to reject that possibility, however, saying that the two systems, although they are "interdigitated in visual areas 1 and 2" (17 and 18), go on to become "segregated into entirely separate areas." Except for the magno projections from area 17 to MT and thence to parietal cortex (whose functional significance we just discussed), other evidence for this statement was not given; and some of the neurological and related data cited in the text here cast doubt on it.

Finally, cortical processing assemblies in general appear to be confined to the upper, or phylogenetically newer layers (II–IV), and these layers, like the "blobs" in area 17, show evidence of differentially high metabolic rates (Pope, 1987). For this reason, I suggested that memories formed at the level of lower III might be more fluid or subject to change over time than those formed at the level of V.

It is then the resultants of such parvocellular processing activity that tend to become "integrated" (reduced to their most commonly encountered or general features) and retained as memories, in large-celled throughput arrays in layers lower III and V. That conclusion is consistent with the finding (reported in chapter 5) that these layers are differentially subject to LTP and show the highest concentrations of phosphate-activated glutaminase.

Considered as a whole—a system—it is these arrangements, including the several shortcuts just described, that put the information conserved qua memory efficiently at the service of present behavior; which is to say, of the evolutionary logic that sets action first.

Appendix I

The report by Corbetta et al. (1990) concerns a complex psychophysical experiment in which Ss fixated gaze on a point in the center of a backlit screen (fixation monitored by EOG), and were required to detect changes (or no change) between patterns displayed in two 400 msec frames separated by 200 msec "off;" 2500 msec were then allowed for a key-press response.

Each of the two displays consisted of a pattern of colored (red or green) rectangles. These moved "horizontally as a coherent sheet either to the left or to the right. The direction of motion was constant within a trial but was randomly shifted across trials. The shape, color and velocity of all elements might independently change between the first and second frames in a trial. The subject's task was to compare the first stimulus frame with the second, and report (by a key-press) if the two frames were the same or different for a particular stimulus dimension, specified at the beginning of each experimental block."

During three blocks of trials, Ss had to make same-or-different discriminations for shape, color and velocity. In each block, for each of these dimensions to be attended to, the same/different ratio was 50/50 (presumably at random). Also, "same and different trials contained (in equal proportions) stimulus changes in zero, one or two of the. . .irrelevant dimensions. For instance, during same and different trials in a color block, velocity and shape might stay constant in both frames, velocity might change, shape might change" or both might change.

(There was also a fourth block of trials in which Ss were not told what to look for, but in which, in any given pair of frames, a randomly picked dimension had a 50/50 chance of being different. Thus all dimensions in both frames might be the same, or, 50% of the time, one of the three would be different. These were called "divided attention" trials. Not surprisingly, discrimination performance was less accurate in these than in the "selective attention" series.)

Cerebral bloodflow (cbf), tracked by positron emission tomography (PET scans) during the selective attention, divided attention and "no task" trials (9 Ss, ages 22–34) produced these areal findings:

(1) In ventromedial occipital cortex, a focus in the lingual gyrus for both color and shape—bilaterally for color, unilaterally on the left for shape. No velocity responses.

(2) In the fusiform and parahippocampal gyri, bilaterally, for shape. (See authors' Fig. 2.)

(3) In a region of (left) mesial cortex just anterior to, or in, the parieto-occipital sulcus, for shape.

(4) Midway along the superior temporal sulcus ("between areas 21 and 22"), bilaterally, for shape. "The magnitude of this response was stronger on the right side. None of these regions was activated by color or velocity."

(5) A strip of cortex, equivalent to dorsolateral 19, was bilaterally activated by (attention directed to) color (changes). (Also reported at the site in lingual gyrus, bilaterally, described in item 1 above).

(6) In the inferior parietal lobule, area 39, left side only, (attention directed to) velocity.

This last region overlaps one that some of these same authors reported elsewhere was involved in a "smooth-pursuit visual tracking task." They add that "these two areas have been proposed as human homologs of the middle superior temporal (MST) and middle temporal (MT) areas in monkeys. The finding that attentional modulation was present in an MST-like area but not in an MT-like area, agrees with single-unit studies in macaques, showing that only MST neurons appear to be affected by 'extraretinal' attention signals."

This distinction is important, since in chapter 2, we concluded that the relative-motion-detection units in MT in the owl monkey represented a system closer to the prime visual receiving areas, and therefore one likely to mature developmentally into a virtually "hard-wired" one. MST neurons represent a further stage in the visual relay and are therefore more likely to be involved in the hippocampal "registration" that forms an important part of attention.

According to the model of memory functions developed here, the parietal representation of spatial relations would develop experientially by cross-modal interaction of visual, auditory and somatosensory data at the cortical level, with an important subcortical component (the vestibular). This primary form of spatial representation has a specific context corresponding to the space in which the body operates.

However, just as interareal relay leads to generalization in the auditory and visual systems, so in distal somatosensory association cortex (e.g., of 7b), the space represented will tend to be abstracted from its original context, and can "stand for" or be associated with anything. This is functionally analogous to the generic visual memory "dog" which stands for, or can be elicited by, the sight of any doglike creature. The difference is that "dog" refers to a class of entities. Space, generalized, refers to a container or abstract coordinate system in which anything can be arranged.

This generalized form of space turns up in our frequent use of spatial terms to describe nonspatial relations. We speak of prices, human types, evolutionary forms or musical tones as "high" or "low," of scales as running "up" or "down," of some lifestyles as "upscale," of some persons as being "highly" regarded, and so on. At one extreme, we have the abstract spaces of mathematics or the space-time manifolds of physics. Toward the other, parietal "space" remains closer to its original body-centered form and its motivational-cum-affective roots in the limbic system. (For the circuitry involved, see Mesulam, 1985, pp. 150–155.)

This motivational-affective component is revealed by our "loaded" use of words such as right and left. Right, being the preferred hand for most of us, translates into "rightness" (of choice, judgment, etc.), righteousness (of conduct) or human rights (meaning privileges we hope are God-given). Left, the nonpreferred hand, translates into "sinister," bar sinister (illegitimate), or the political Left (always bad). The same sentiment turns up idiomatically in expressions such as "That was a left-handed way of doing things."

Caramazza and Hillis (1990) reported the case of a 77-year-old left-handed woman who, after a stroke involving the left parietal white matter and an area "adjacent to the head of the caudate nucleus," developed a disorder that selectively affected this abstract space. It is discussed in chapter 2 here.

Appendix II

Tranel and Damasio (1985) report two cases of prosopagnosia in which the subjects could not verbally or consciously identify 8 once-familiar faces from photographs randomly intermixed with photos of 42 unfamiliar faces. The same subjects showed statistically significant electrical skin conductance (SCR) responses to the familiar faces.

Subject 1 was a 62-year-old right-handed woman who had suffered bilateral strokes involving "the white matter of both occipitotemporal regions" but sparing "the most mesial and rostral regions of the inferior visual association cortex. As determined by a comprehensive neuropsychological assessment, her intellect, language and visual perception are normal." She had both retrograde and anterograde prosopagnosia, being unable to recognize any faces, including her own, seen before or after onset of her condition seven years previously.

Subject 2 was a 20-year-old right-handed woman who had bilateral occipitotemporal lesions resulting from herpes simplex encephalitis. "The lesions are located more anteriorly than in Subject 1, so they also spare the mesial and rostral aspects of the inferior visual association cortex. Her language abilities are intact and her visual perception is compatible with normal recognition of faces learned before onset of the illness."

This patient was characterized as having a "lack of retrograde prosopagnosia." She did, however, have the anterograde form, making her unable to make visual identification of faces seen after onset of her illness.

Whereas S1 showed significant SCRs to familiar faces (of family or public figures) without being able to identify any of them, S2 gave galvanic skin responses both to faces she recognized (because they had become known to her before her illness) and also to those she did not (because she had encountered them after her illness).

In other words, S1 could neither access long-term visual memories for faces nor form new ones; whereas S2 could not form new ones but could access those already in being.

These findings suggest that the more anterior lesion in S2 selectively damaged visual pathways afferent to (but not those efferent from) entorhinalis, the subiculum, and the FD and hippocampus. In S1 both sets of pathways appear to have been interrupted.

The SCR shown by these patients is perhaps part of the response pattern defined in the text as "registration." That is, it reflects a "blip" or transient rise in central arousal such as tends to occur in "orienting" or "novelty" responses. Occurring at the start of a stimulus train, it leads to the "primacy" effect. That is, it causes the first item in the train to be "registered" as a result of the accompanying rise in hippocampal activity.

The fact that Henry showed no primacy effect indicates that the hippocampus figures in it, and that registration amounts to an increase in the probability that the item registered will either (a) have a longer-than-average duration in STM or (b) subsequently be incorporated (consolidated) into LTM.

In S1 and S2, the SCR suggests an orienting response, triggered by the "novelty" effect of a familiar face suddenly appearing in a string of unfamiliar ones—a type of "registration" that perhaps figures normally in specific recognition when that occurs under unusual circumstances, or as a surprise. Moreover, the fact that S2 exhibited this "surprise" response to faces she could also identify, suggests that in her (and possibly in S1 as well) the hippocampus was intact.

They both showed signs of "registering" familiar visual inputs. But we must then ask by what routes (especially in S1) was the hippocampal system informed that the input was in fact familiar? The vertical organization of memory functions proposed in the text (chapter 7) may provide an answer to this question.

For example, may visual data representing a familiar face carry brainstem motivational or affective tags that are bidirectionally connected, e.g., via the mammillary bodies, with correspondingly tagged memories at the level of the cingulate or of entorhinalis and the "subicular complex?" (Swanson, 1987.) Is it via these pathways, and the amygdala, that visual recognition results in an arousal response whose autonomic index, in these patients, was a change in skin conductance?

In other words, were these patients showing the type of motivationally or affectively driven retrieval described in chapter 7—a retrieval in which the only missing piece was conscious awareness that what had been retrieved was, in fact, familiar? The hypothesis developed here would say that the missing piece was the white matter connections and/or their cortical areas of origin that had been lesioned, thereby cutting off the "core" components of a given long-term visual memory from its cortical referents. The faces could be generically but not specifically identified in what we call "consciousness."

This, in turn, relates to the question raised in chapter 8, part 2, as to whether consciousness itself may not be a function primarily of posterior cortex, and primarily an outgrowth of long-term memory. (See the discussion of the "boundaries" of consciousness, e.g., as indicated by the phenomenon of priming.)

Appendix III

In the current language of neurocomputing, the inhibitory phase of cortical activation sets up a system of energy "valleys." In the ensuing facilitatory phase, energy then flows into these from the highest nearby "hills" (e.g., from foci representing concurrent or most recent sensory input). This would correspond to what Hopfield calls the energy minimum-seeking property of the network. He is speaking of theoretical or simulated neural networks, but the property may be that of real ones (Hopfield, 1988; see also Tank and Hopfield, 1987).

It is of some interest that Little and Shaw (1988) find that in their network simulating memory functions, "there is some optimum value for v or the noise as far as the storage capacity of the network is concerned. Having some noise serves a useful purpose by increasing the storage capacity over that of a comparable network which is free of such statistical fluctuations."

This dependence upon some optimal noise level, above or below which the network does not function up to its full potential, has an evident parallel in the CNS. The fact that reticulocortical activation is nonspecific and has been shown to convey a minimum of information qualifies it as, within limits, a source of useful noise. Extremely high and extremely low levels of central activation seem, in fact (though in characteristically different ways) to impair learning or other sorts of performance. According to Denenberg, behavioral efficiency is an inverted U function of the degree of central activation (Denenberg, 1967).

Memory formation, itself a residue of central nervous work (as represented, for example, by LTP and the differential modifications of cell structure discussed in the text), has a longer-term energy conserving feature. As suggested by this author (Fair, 1965), it acts to decrease the amounts of central nervous work elicited by day-to-day sensory experience.

For exactly in proportion as memory serves to make a wider and wider range of sensory inputs immediately recognizable, the probability that any given input will touch off "orienting" or "novelty" responses is reduced. This is no more than a restatement of Sokolov's principle, according to which cortical memory systems tend to damp back "orienting" responses if the inputs triggering them at the brainstem level are in fact familiar and nonsignificant (Sokolov, 1960). In other words, memory—familiarity with the environment—reduces the number of episodes of central activation and fearlike overresponse arising out of sheer uncertainty.

Appendix IV

The patient, in the study by Halligan and Marshall (1991) was a 57-year-old right-handed male who, following a stroke, was admitted with a severe left hemiparesis and left homonymous quadrantanopia. "There was a full range of ocular movements" and normal visual acuity.

Right-sided damage was extensive, including the pons, cerebral peduncle, internal capsule, and "most of the lateral and medial temporal cortex." The frontal pole and frontal cortex, except for the right lateral inferior aspect, were spared, as was occipital cortex (though apparently there was some interruption of the right optic tract).

The patient did not show personal neglect—could touch body parts on his left side with his right hand, with eyes either open or closed. However, when asked to bisect a line, he shifted the midpoint rightward. But when the line was shown on a screen 2.88 meters from the patient, and he used a light pen to bisect it, he did so with significantly greater accuracy and with reduced scatter (variance).

It is interesting that the table the authors used to summarize these results shows that for five out of the six line lengths used, the patient showed a (probably not significant) leftward shift in his errors. (His errors at close range were all shifted rightward.)

Halligan and Marshall describe this as a case of "classical left visuo-spatial neglect within peripersonal space." In other words, spatial representation in the brain can take three forms—the "literal bodily" form, the space-within-reach form, and the largely (but never totally) self-unrelated form. In human infants or baby macaques, can spatial awareness be shown to follow this sequence developmentally? Or do the last two forms of space perception develop, at different rates, on parallel but separate tracks?

(Everyday parental observation suggests the latter. Small children appear to learn more about space from the purely visual side than at first they are able to handle—hence they have to be protected from falling down stairs.)

Note added in press. Duhamel et al. (1992) report that single units in "lateral intraparietal cortex" in the macaque have retinal receptive fields that shift as the eyes move. The responses of these neurons thus correspond not to particular areas in the retina but to parts of the animal's visual space. In these experiments, the animals were trained to fixate a target. A second point, called the stimulus, was then illuminated nearby. Parietal units with corresponding receptive fields responded to the stimulus at a latency of about 70 msec. This event appears to set up a transient memory or residual facilitation, such that if, in a next experiment, the fixation point lies outside of the visual field of the parietal neuron and is then shifted back into it, the animal makes a saccade. As that occurs, the parietal neuron is activated <80 msec *in advance* of the beginning of the saccade, or about 150 msec before the appearance of the stimulus in its visual field. As the authors note, these findings suggest a clear basis for the perceived constancy of visual space, as well as an intentional or "hold" mechanism by which we tag items in our visual space for possible further attention.

In terms of the principles developed in this monograph; the interareal relay of sensory data in parietal cortex generalizes from vestibular, somatosensory and visual inputs so as to extract visual space as an invariant common to much of our experience. There are even gradations in the generality of this space, reflecting the several stages in the relay by which it is established. If this account is correct, it should permit systematic correlation between the locus and amount of parietal damage, and the disorders in space perception that result.

References and Notes

Abt JP, Essman WB, Jarvik ME (1961): *Science* 133:1477–1478

Akiyama H et al. (1989): *Soc Neurosc Abstr* 15(1):327.7

Alkon D (1984): *Science* 226:1037–1045

Alkon D (1989): *Scientific American* 261(1)(July):42–50

Allman J (1987): In *Encyclopedia of Neuroscience*. G Adelman, ed. Birkhäuser Boston, 2:1283–1287

Ambros-Ingerson J, Granger R, Lynch G (1990): *Science* 247:1344–1348

Anderson JR (1976): *Language, Memory and Thought*. Erlbaum, Hillsdale NJ

Anderson MF, Winterson BJ (1988): *Soc Neurosc Abstr* 14(2):319.4

Angevine JB (1965): *Experimental Neurology*. Supplement 2:1–70

Antelman SM et al. (1988): *Ann NY Acad Sc* 537:262–271

Aston-Jones G et al. (1986): *Science* 234:734–738

Ayer-LeLievre C et al. (1988): *Science* 240:1339–1341

Bailey CH, Chen M (1983): *Science* 220:91–93

Bailey P, Von Bonin G (1951): *The Isocortex of Man*. University of Illinois Press

Bayer SA, Yackel JW, Puri PS (1982): *Science* 216:890–892

Bear M, Cooper LN, Ebner FF (1989): In *Neuroscience Year*. G Adelman, ed. Birkhäuser Boston, pp. 156–160

Beaulieu N, Morris R, Petrides M (1988): *Soc Neurosc Abstr* 14(2):494.10

Bekkers JM, Stevens CF *Nature* 346:724–729

Benowitz LI, Perrone-Bizzozero NI (1989): In *Neuroscience Year*. G Adelman, ed. Birkhäuser Boston, pp. 118–120

Beracochea DJ, Jaffard R (1988): *Soc Neurosc Abstr* 14(2):494.26

Bergman H, Wichmann T, DeLong MR (1990): *Science* 249:1436–1438

Berlucchi G, Sprague JM (1981): In *The Organization of the Cerebral Cortex*. FO Schmitt et al., eds. MIT Press, pp. 395–414

Bindman LJ, Murphy KPSJ (1988): *Soc Neurosc Abstr* 14(1):226.1. For a study of NMDA-receptor mediated LTP, involving units of layers III, V and VI in neocortx, see Artola A and Singer W (1987): *Nature* 330:649–652

Bishop G (1958): In *Reticular Formation of the Brain.* HH Jasper et al. eds. Little Brown, pp. 413–421

Blakemore C (1974): In *The Neurosciences Third Study Program.* FO Schmitt et al., eds. MIT Press, p. 112. Pribram (personal communication, 1989) reports that in his laboratory "we have . . . shown with micro and macro electrode procedures that cells in the striate cortex respond to the results of experience (learning) according to Blakemore's suggestion."

Bloedel JR (1987): *Science* 238:1728–1729

Bloom FE (1981): In *The Organization of the Cerebral Cortex.* FO Schmitt et al., eds. MIT Press, pp. 359–370

Brown TH, Chapman PF, Kairiss EW, Keenan CL (1988): *Science* 242:724–728

Burckhardt G (1890–1891): *Allg Z Psychiat* 47:463–548

Caramazza A, Hillis AE (1990): *Nature* 346:267–269

Caramazza A, Hillis AE (1991): *Nature* 349: 788–790

Casey KL, Jones EG (1980): *Neurosc Res Prog Bull* 16(1):103–118

Chamberlain TJ, Halick P, Gerard RW (1963): *J Neurophysiol* 26:662–673

Chapman PF, Brown TH (1988): *Soc Neurosc Abstr* 14(I):566

Chernevskaya NI, Obukhov AG, Krishtal OA (1991): *Nature* 349:418–420 These findings are interesting in that calcium N-channel block is reportedly irreversible when mediated by glutamate or aspartate, but reversible when adenosine is also present.

They add that "this reversible effect is accompanied by a long-lasting change in excitatory transmission pharmacology and may be related to plasticity." (The transmission referred to is between CA3 and CA1 cells.) "We suggest that a regulatory complex exists which controls N-channels through functionally interconnected NMDA and adenosine receptors."

Cole AJ, Saffen DW, Baraban JM, Worley PF (1989): *Nature* 340:474–476

Colonnier M (1981): In *The Organization of the Cerebral Cortex.* FO Schmitt et al., eds. MIT Press, p. 141

Cooper L (1981): In *The Organization of the Cerebral Cortex.* FO Schmitt et al., eds. MIT Press, pp. 479–503

Cooper JR, Bloom FE, Roth RH (1986): *The Biochemical Basis of Neuropharmacology.* 5th Ed. (paperback), Oxford, pp. 255–256 and 263

Corbetta M et al. (1990): *Science* 428:1556–1559

Corkin S, Sullivan EV, Gabrieli JDE, Yucaitis J (1989): *Soc Neurosc Abstr* 15(1):5.4

Cotman CW, Nieto-Sampedro M (1985): In *Neuroscience.* PH Abelson et al., eds. Amer Assoc Adv Sc, p. 76

Coyle JT, Price DL, De Long MR (1985): In *Neuroscience*. PH Abelson et al., eds. Amer Assoc Adv Sc, pp. 418–431

Crawley JN (1990): *Concepts in Neurosc* 1(2):287–298

Creutzfeldt OD, Heggelund P (1975): *Science* 188:1025–1027

Crutcher KA, Collins F (1982): *Science* 217:67–68

Damasio AR (1985): In *Principles of Behavioral Neurology*. M-M Mesulam, ed. FA Davis, pp. 262–266

Davies SN, Lester RAJ, Reymann KG, Collingridge GL (1989): *Nature* 338:500–503

Davies SN, Starkey SJ, Pozza MF, Collingridge GL (1991): *Nature* 349:609

Decker MW, McGaugh JL (1988): *Soc Neurosc Abstr* 14(1):288.7

De Kloet ER et al. (1987): *Ann NY Acad Sc* 512:351–361

DeLong MR (1987): In *Encyclopedia of Neuroscience*. G Adelman, ed. Birkhäuser Boston, pp. 113–115

Denenberg VH (1967): In *Neurophysiology and Emotion*. DC Glass, ed. Rockefeller University Press, p. 172 ff

Desmedt J (1981): In *The Organization of the Cerebral Cortex*. FO Schmitt et al., eds. pp. 441–473

de Toledo-Morrell et al. (1991): *Arch Neurol* 48:605–609. These authors note the role of the hippocampus in "registration" and add that "there is now a substantial body of work indicating that the P300 potential may be generated in or near the hippocampus."

In six elderly patients (mean age, 71) showing apparent early signs of Alzheimer's disease, they found that the latency of the P-300 did not differ signicantly from that in age-matched controls. However, if the number of target letters in a short-term memory test was increased from 1 to 5, the P-300 tended to disappear with this increase in "load." The accuracy of the patients' short-term recall, vs. controls, was less affected. How might these patients have done on tests of primacy and recency effects in serial learning?

Deweer B, Sara SJ (1985): *Ann NY Acad Sc* 444:507–509

De Wied D, De Kloet ER (1987): *Ann NY Acad Sc* 512:328

Duhamel, J-R, Colby CL, Goldberg ME (1992): *Science* 255:90–92

Dumont JPC, Robertson RM (1986): *Science* 233:849–853

Eimas PD (1985): *Scientific American* 252(1)(January):46–52

Emson PC, Hunt SP (1981): In *The Organization of the Cerebral Cortex*. FO Schmitt et al., eds. MIT Press, pp. 325–345

Evinger C, Pellegrini JJ, Manning KA (1989): *Ann NY Acad Sc* 563:87–100

Fair CM (1965): *Neurosc Res Prog Bull* 3(1):27–62

Fair CM (1984): This paper (unpublished; synopsized in Fair, 1988) proposed that the brain is a system that uses the energy available to it to reduce its own entropy, essentially by changes in connectivity that increase the organization of its data-on-file. These changes were inferred to result from the brain's continuous electrical activity, and to occur between, as well as during, occasions of overt behavior or of attention to particular inputs or tasks. The effect of that activity was to increase the amount of information, as opposed to factual data, held in memory—information being defined as negentropic. Shannonian information was then its inverse, or the corresponding cost in central nervous work. By appropriate substitutions, the paper proposed to use equations from classical thermodynamics as a way of estimating mental efficiency. It defined intelligence as in part proportional to the brain's work capacity per unit time, a power factor not necessarily equivalent to its ability to establish useful new connections among the data available to it.

Fair CM (1987): *Science* 238:1730–1731

Fair CM (1988): *Memory and Central Nervous Organization*. New York: Paragon House

Fair CM (1989): *Language Sciences* 11(4):355–366

Fallon JH, Loughlin SE (1987): Monoamine innervation of cerebral cortex and a theory of the role of monoamines in cerebral cortex and basal ganglia. In *Cerebral Cortex*. Vol. 6, EG Jones and A Peters, eds., Plenum

Flexner JB, Flexner LB, Stellar E (1963): *Science* 141:57–59

Fox K, Daw N, Sato H, Czepita D (1991): *Nature* 350:342–344

Friedberg M (1990): Results presented in seminar, at Brown University (graduate thesis material). The histological data showing the extent of NMDA-mediated laminar ablation were quite clear.

Fulton JF (1951): *Physiology of the Nervous System*. Oxford University Press, pp. 288–330

Fulton JF (1951a): *Frontal Lobotomy and Affective Behavior*. Norton, p. 21

Funahashi S, Bruce CJ, Goldman-Rakic PS (1989): *J Neurophysiol* 61(2):331–342

Fuster JM, Jervey JP (1981): *Science* 212:952–955

Fuxe K et al. (1987): *Ann NY Acad Sc* 512:362–393. See especially Table 3, p. 376.

Gabriel M, Foster K, Orona E (1980): *Science* 208:1050–1052

Gastaut H (1958): In *Reticular Formation of the Brain*. HH Jasper et al., eds. Little Brown, pp. 576–577

Gentile CG et al. (1988): *Soc Neurosc Abstr* 14(2):494.12

Geschwind N (1974): *Selected Papers on Language and the Brain*. RS Cohen and MW Wartofsky, eds. p. 99 Reidel

Gilbert C, Wiesel T (1981): In *The Organization of the Cerebral Cortex*. FO Schmitt et al., eds. MITPress, p. 171

Goh JW, Pennefather PS (1989): *Science* 244:980–983

Goldman-Rakic PS (1986): Lecture delivered at Nauta Symposium MIT, May 27 See also Funahashi et al., above. The former report includes the fronto-parietal connections discussed in the text here. Both reports illustrate the frontal "hold" functions or working memory mechanism also discussed here.

A third study (PS Goldman-Rakic et al. (1989): *Proc Nat Acad Sc* 86:9015) reports the convergence of a symmetrical DA synapse and an asymmetrical (possibly glutamate) terminal on the same dendrite of a layer II cell in macaque prefrontal cortex. This finding may illustrate, at the unitary level, one form of the catecholaminergic inhibitory "reset" mechanism mentioned passim in the text.

Gray CM, Konig P, Engel AK, Singer W (1989): *Nature* 338:334–337

Graybiel AM (1974): In *The Neurosciences Third Study Program.* FO Schmitt et al. eds. MIT Press, pp. 205–241

Graybiel AM, Berson DM (1981): In *The Organization of the Cerebral Cortex.* MIT Press, pp. 285–319

Gross CG, Bender DB, Rocha-Miranda CE (1969): *Science* 166:1303–1306 Re unitary responses in macaque inferotemporal cortex, see these authors, 1974. In *The Neurosciences Third Study Program,* FO Schmitt et al., eds. MIT Press, pp. 229–238

Grover LM, Teyler TJ (1990): *Nature* 347:477–479

Guldin WO, Akbarian S, Grusser O-J (1989): *Soc Neurosc Abstr* 15:212–14

Halligan PW, Marshall JC (1991): *Nature* 350:498–500. See note 4, added in press.

Harootunian AT, Kao JPY, Paranjape S, Tsien RY (1991): *Science* 251:75–78

Heath RG (1964): In *The Role of Pleasure in Behavior* RG Heath, ed. Harper and Row

Hebb DO (1949): *The Organization of Behavior.* Wiley

Heimer L, Alheid GF, Zaborsky L (1989): "Basal Forebrain and Substantia Innominata." In Supplement 1, *Encyclopedia of Neuroscience,* G Adelman, ed. Birkhäuser Boston

Henneman E (1991): *Ann NY Acad Sc* 627:165–168

Herrick CJ (1948): *The Brain of the Tiger Salamander.* University of Chicago Press

Hitchcock JM, Sananes CB, Davis M (1988): *Soc Neurosc Abstr* 14(2):494.9

Hockfield S (1987): *Science* 237:67

Hökfelt T et al. (1985): In *Neuroscience.* P Abelson et al., eds. Amer Assoc Adv Sc, pp. 199–215

Hopfield JJ (1988): In *Neurocomputing.* JA Anderson and E Rosenfeld, eds. MIT Press, pp. 580–583. See also Tank DW and Hopfield JJ (1987): *Scientific American* 257(6):104–114

Hopkins WF, Johnston D (1984): *Science* 226:350–352

Horridge GA (1964): In *Comparative Neurochemistry.* D Richter, ed. pp. 35–47 Pergamon, New York

Hubel D, Wiesel T (1962): *J Physiol* 160:106–154 (London)

Hubel D, Wiesel T (1963): *J Physiol* 29:994–1002

Hubel D, Wiesel T (1974): *J CompNeurol* 158:295–306

Hubel D (1989): Personal communication

Huszagh VA, Infante JP (1989): *Nature* 338:109

Hydén H, Lange PW (1968): *Science* 159:1370–1373 Also 1970, in *The Neurosciences: Second Study Program*. FO Schmitt et al, eds. Rockefeller University Press, pp. 278–289

Iriki A, Pavlides C, Keller A, H Asanuma (1989): *Science* 245:1385–1387. See also (1989): *Soc Neurosc Abstr* 15(1):113.4

Iversen S (1986): Data presented at Nauta Symposium MIT, May 27

Izquierdo I, Netto CA (1985): *Ann NY Acad Sc* 444:162–177

Jacob F (1977): Evolution and Tinkering. *Science* 196:1161–1166

Jacobs BL (1986): *Ann NY Acad Sc* 473:70–77

Janowsky JS, Shimamura AP, Kritchevsky M, Squire LR (1989): *Behavioral Neurosc* 103(3):548–560

Jasper HH (1958): In *Reticular Formation of the Brain*. HH Jasper et al., eds., p. 319 ff

Jones EG (1974): The anatomy of extrageniculostriate visual mechanisms. In *The Neurosciences: Second Study Program*. FO Schmitt et al. eds., MIT Press, pp. 215–228

Jones EG (1981): In *The Organization of the Cerebral Cortex*. FO Schmitt et al., eds. MIT Press, pp. 199–235

Kaas JH, Nelson RJ, Sur M, Merzenich MM (1981): In *The Organization of the Cerebral Cortex*. FO Schmitt et al., eds. MIT Press, pp. 237–262

Kaas JH, Krubitzer LA, Chino YM, Langston AL, Polley EP, Blair N (1990): *Science* 248:229–231. See also Calford MB, and Tweedale R (1990): *Science* 249:805–807. These authors report that in the macaque, reversible partial denervation (anesthesia) of a digit causes corresponding receptive field changes (expansions) at mirror foci, bilaterally, in 3b. The result is puzzling, since ipsilateral receptive fields have mostly not been found in primary somatosensory cortex, and callosal connections between distal parts of the forelimb "are thought to be sparse." Receptive fields returned to normal in ~30 minutes, suggesting return of the local inhibition that normally defines them. In the more extensive deafferentation used by Kass et al. (1981), somatosensory units became unresponsive for several weeks before developing new receptive fields.

Kandel E, Schwartz JH (1985): In *Neuroscience*. PH-Abelson et al., eds. Amer Assoc Adv Sci pp. 381–402

Keene JJ (1975): *Exper Neurol* 49:97–114

Keinanen K et al. (1990): *Science* 249:556–560

Keller A, Iriki A, Asanuma H (1989): *Soc Neurosc Abstr* 15(1):113.3

Kesner RP, Novak JM (1982): *Science* 218:173–175 8 Oct

Keverne EB, Brennan P (1990): *Science* 250:1223–1226

Konishi M (1986): Quoted in *Science* 231:1068

Kosofsky BE, Kowall NW (1989): *Soc Neurosc Abstr* 15(1):6.1

Krieger DT (1985): In *Neuroscience.* P Abelson et al., eds. Amer Assoc Adv Sc, pp. 309–331

Kristeva R (1984): *Ann NY Acad Sc* 425:477–482

Kromer LF (1987): *Science* 235:214–216

Lilly J (1960): In *The Central Nervous System and Behavior.* Transactions of the Third Conference, Josiah Macy Jr Foundation, p. 51

Lisberger SG (1984): *Science* 225:74–76

Little WA, Shaw GL (1988): In *Neurocomputing.* JA Anderson and E Rosenfeld, eds. MIT Press, p. 238

Livingstone M, Hubel D (1988): *Science* 240:740–749, 6 May

Llinás RR (1987): In *Encyclopedia of Neuroscience.* G Adelman, ed. Birkhäuser Boston, pp. 201–203

Lorente de Nó R (1934): *J f Psychol und Neurol* Bd 46, Heft 2 u 3, pp. 113–176

Lorente de Nó R (1951): In *Physiology of the Nervous System* by JF Fulton, 3rd Edition. Oxford University Press, pp. 288–330. Original date of paper, 1938.

Lund JS (1981): In *The Organization of the Cerebral Cortex.* FO Schmitt et al., eds. MIT Press, pp. 105–124

Luria AR (1976): *The Neuropsychology of Memory.* Washington DC: Winston, p. 5

Lynch G, Baudry M (1985): In *Neuroscience.* P Abelson et al., eds. Amer Assoc Adv Sci, pp. 403–431

Malenka RC et al. (1989): *Nature* 340:554–557 (17 August)

Malinow R, Schulman H, Tsien RW (1989): *Science* 245:862–866

Malinow R, Tsien RW (1990): *Nature* 346:177–180 ,

Malinow R (1991): *Science* 252:722–724 See also Manabe T, Renner P, and Nicoll RA (1992): *Nature* 355:50–55

Maren S, Cox A, Gabriel M (1989): *Soc Neurosc Abstr* 15(1):37.18

Mason JW (1958): In *Reticular Formation of the Brain.* HH Jasper et al., eds. Little Brown, pp. 645–662. In the macaque, Mason found that very steep rises in corticosteroid plasma levels could be elicited by electrical stimulation of several sites in the amygdala. By contrast, hippocampal stimulation, over 90 minutes, produced a slight rise in plasma levels, followed, at 24 and 48 hours, by a marked lowering Bilateral hippocampectomy or bilateral fornix section blocked the normal diurnal (PM) fall in plasma corticosteroid levels in these animals. These findings suggest a strong negative feedback action, possibly (as suggested by McEwen et al., 1987) via hippocampal or subicular projections to the paraventricular nucleus.

Masu M et al. (1991): *Nature* 349:760–765

Matthews PBC (1988): *Encyclopedia of Neuroscience*. G Adelman, ed. Birkhäuser Boston, 2:710–712

McEwen B, Chao H, Spencer R, Brinton R, MacIsaac A, Harrelson A (1987): *Ann NY Acad Sc* 512:394–401

McGaugh JL (1985): *Ann NY Acad Sc* 444:150–161

Mesulam, M-M (1985): In *Principles of Behavioral Neurology*. M-M Mesulam, ed. FA Davis, p. 129

Milner B (1964): Talk delivered in seminar, Department of Psychology MIT, Oct 21.

Miserendino MJD, Sananes CB, Melia KR, Davis M (1990): *Nature* 345:716–718

Montarolo PG, Goelet P, Castellucci VE, Morgan J, Kandel E, Schachter S (1986): *Science* 234:1249–1254

Moore RY (1977): *Neurosc Res Prog Bull* 15(2):160–168

Mott DD, Lewis DV (1991): *Science* 252:1718–1720

Mountcastle VB (1986): Data presented at the Nauta Symposium MIT, May 28

Mouradian RD, Sessler FM, Waterhouse BD (1988): *Soc Neurosc Abstr* 14(1): 52.11

Müller W, Connor JA (1991): *Nature* 354:73–76

Murray EA, Mishkin M (1985): *Science* 228:604–606

Nauta WJH (1982): In *Advances in Neurology*. Vol 35, AJ Friedhof and TN Chase, eds. Raven Press, pp. 41–47. This is an important paper concerning innervation of various parts of the striatum (and of n. accumbens and the anterior peforated substance) from the cingulate, amygdala, hippocampus and ventral tegmental area.

Nauta WJH, Feirtag M (1986): *Fundamental Neuroanatomy*. Freeman, pp. 124 and 241

Nelson PG, Chang Yu, Fields RD, Neale EA (1989): *Science* 244:585–587

Nielsen JM (1946): *Agnosia, Apraxia and Aphasia: Their Value in Cerebral Localization*. 2nd ed. New York: Hoeber

Ojemann G, Mateer C (1978): *Science* 205:1401–1403

Olds J (1975): In *The Neurosciences*. FG Worden et al., eds. MIT Press, pp. 375–400. Cf. also Olds ME and Olds J (1963): *J Comp Neurol* 120(2):259–295

Pechtel C, Masserman JH (1954): *Ann NY Acad Sc* 58(2):256–260

Penrose R (1989): *The Emperor's New Mind*. Oxford University Press, pp. 404–449

Pert C (1980): *Neurosc Res Prog Bull* 16(1):133–141

Peterson LR (1966): *Scientific American* 215(July):90–95

Peterson SL (1985): *Abstr No. 245.14* Annual Meeting, Soc Neurosc, Dallas

Phillips HS, Hains JM, Laramee GR, Rosenthal A, Winslow JW (1990): *Science* 250:290–294

Pillon P, Bakchine S, Lhermitte F (1987): *Arch Neurol 44*:1257–1262

Poggio T, Girosi F (1990): *Science* 247:978–982 See Figs. 1 and 2, showing their mathematically derived invariance-extracting networks.

Poggio T, Torre V (1981): "A Theory of Synaptic Interactions" in *Theoretical Approaches in Neurobiology.* WE Reichardt and T Poggio, eds. MIT Press, pp. 28–38

Pons TP, Garraghty PE, Friedman DP, Mishkin M (1987): *Science* 237:417–420

Pons TP et al. (1991): *Science* 252:1857–1860

Pope A (1987): In *Encyclopedia of Neuroscience.* G Adelman, ed. Birkhäuser Boston, 1:211–214

Pribram K, Kruger L (1954): *Ann NY Acad Sc* 58(2):109–138

Pribram K (1974): In *The Neurosciences Third Study Program.* FO Schmitt et al., eds. MIT Press, pp. 249–261

Pribram K, McGuinness D 19:75 *Psychol Rev* 82(2):116–149

Pribram K 1984; (1989): Personal communications

Price JL (1987): In *Encyclopedia of Neuroscience.* G Adelman, ed. Birkhäuser Boston, 1:40–42

Quirk GJ, Muller RU (1989): *Soc Neurosc Abstr* 15(1):163.21

Rakic P et al. (1986): *Science* 232:232–234

Rasmussen H (1989): *Scientific American* 261(4)(Oct):66–73

Routtenberg A (1978): *Scientific American* 239(5):154–164

Rubin DC (1989): In *Neuroscience Year* G Adelman, ed. Birkhäuser Boston, p. 101

Rusak B, Robertson HA, Wisen W, Hunt SP (1990): *Science* 248:1237–1240

Sacks O (1987): *The Man Who Mistook His Wife for a Hat.* Perennial Library, p. 113

Sapolsky RM, Pulsinelli WA (1985): *Science* 229:1397–1400

Sara SJ (1985): *Ann NY Acad Sc* 444:178–193

Sara SJ, Dyon-Laurent C, Maho C (1988): *Soc Neurosc Abstr* 14(1):288.11

Sarnat HB (1987): *Amer J Dis Children* 41(9):969–980

Sarnat HB, Netsky MG (1981): *Evolution of the Nervous System.* 2nd edition, pp. 340–341

Sartre J-P (1948): *The Psychology of Imagination.* Philosophical Library, p. 127. An academic friend of Sartre's used to ask his students if they could form a mental image of the Parthenon. When any said they could, he would then ask them to count the columns. None could, and many declined "even to try."

The point is that long-term memory, being composed of generalized components, has low resolution; that is, just sufficient resolving power to serve adaptive needs. We can, of course, by changing the scale of our attention, commit patterns of details to memory; but again, how far can we resolve those? Even "photographic" memory tends to be grainy.

Sass KJ et al. (1991): *Arch Neurol* 48(1):48–52

Scheibel A (1987): In *Encyclopedia of Neuroscience*. G Adelman ed. Birkhäuser Boston, 2:1058

Schwanzel-Fukuda M, Pfaff DW (1989): *Nature* 338:161–164

Scoville WB, Milner B (1957): *J Neurol, Neurosurg, and Psychiat* 20:11–21

Shaffer D (1986): In *Anxiety Disorders of Childhood*. Guilford Press, pp. 157–167

Shallice T (1988): *From Neuropsychology to Mental Structure*. Cambridge University Press, p. 297 ff. See also McCarthy RA and Warrington EK (1988): *Nature* 334:428–430

Shimamure AP, Squire LR (1986): *Behavioral Neurosc* 100:165–170

Signoret J-L (1985): In *Principles of Behavioral Neurology*. M-M Mesulam, ed. FA Davis, pp. 169–178

Silva LR, Amitai Y, Connors BW (1991): *Science* 251:432–435

Singer W (1990): *Concepts in Neurosc* 1(1):1–26

Sokolov EN (1960): Neuronal models and the orienting reflex. In *The Central Nervous System and Behavior*. Josiah Macy Jr Foundation, pp. 187–276

Sommer B et al. (1990): *Science* 249:1580–1585

Spinelli DN et al. (1972): *Exp Brain Res* 15:289–304

Squire LR (1986): *Science* 232:1612–1614

Squire LR (1987): *Memory and Brain*. Oxford. See also Press GA, Amaral DG, and Squire LR (1989): *Nature* 341:54–57. In this study, magnetic resonance imaging showed that three amnesic patients, vs. 4 controls, had significant reductions in size of FD, the hippocampus and the subiculum. One of the patients also showed reduction of the parahippocampal gyrus. Bilaterally, the area of the temporal lobe (seen in frontal section) "was nearly identical" in patients and controls.

Stein L (1977): *Neurosc Res Prog Bull* 15(2):172–182

Stryer L (1981): *Biochemistry*. 2nd Ed, Freeman, p. 854

Sutherland NS (1990): *Nature* 347:239

Swanson LW (1987): In *Encyclopedia of Neuroscience*. G Adelman, ed. Birkhäuser Boston, 589–591

Swanson LW, Cowan WM (1975): *Science* 189:303–304

Sweet WH (1980): *Neurosc Res Prog Bull* 16(1):165–166

Talbot JD et al. (1991): *Science* 251:1355–1358

Terzian H, Ore GD (1955): *Neurology* 5(6):373–380

Thompson RF (1986): *Science* 233:941–947

Thompson R (1960): *Science* 132:1551

Tranel D, Damasio AR (1985): *Science* 228:1453–1454

Tulving E, Schacter DL (1990): *Science* 247:301–306

Vaney DI (1986): *Science* 233:444–446

Van Hoesen GW, Pandya D, Butters N (1972): *Science* 175:1471–1473

Van Hoesen GW (1985): *Ann NY Acad Sc* 444:97–112

Vogt BA, Rosene DL, Pandya DN (1979): *Science* 204:205–207

Wahler RG, Hahn DM (1986): In *Child Health Behavior.* NA Krasnegor, JD Arasteh, and MF Cataldo, eds. Wiley-Interscience, pp. 147–167

Waterhouse BD, Sessler FM, Mouradian RD (1988): *Soc Neurosc Abstr* 14(1): 52.11 and 52.12

Weichselgartner E, Sperling G (1987): *Science* 238:778–780

Weinberger NM, Ashe JH, Metherate R, McKenna TM, Diamond DM, J Bakin (1990): *Concepts in Neurosci* 1(1):91–132

Weingartner H et al. (1981): *Science* 211:601–603

Weintraub S, Mesulam M-M (1985): In *Principles of Behavioral Neurology.* M-M Mesulam, ed. FA Davis, p. 83

White EL (1989): Cortical Circuits, Birkhäuser Boston, pp. 26–27

Whitehead AN (1948): *Essays in Science and Philosophy.* Philosophical Library, p. 70

Wilson M, DeBauche BA (1981): *Neuropsychologia* 19:29–41

Winograd T (1975): In *Representation and Understanding: Studies in Cognitive Science.* D Bobrow and A Collins, eds. Academic Press, pp. 185–210

Wright MJ, Ikeda H (1974): In *The Neurosciences: Third Study Program.* FO Schmitt et al., eds. MIT Press, pp. 115–122

Yang, X-D, Korn H, Faber DS (1990): *Nature* 348:542–545

Zalutsky RA, Nicoll RA (1990): *Science* 248:1619–1624

Zimmerman EA (1987): In *Encyclopedia of Neuroscience.* G Adelman, ed. Birkhäuser Boston , 1:55

Zola-Morgan SM, Squire LR (1990): *Science* 250:288–290 1990a *Ann NY Acad Sc* 608:434–456

Index